Electromagnetic Field Standards and Exposure Systems

The SciTech Series on Electromagnetic Compatibility

Mission: This book series provides a continuously growing body of knowledge by incorporating underlying principles with best practices in electromagnetic compatibility engineering. It reflects modern developments of the last quarter century for an ever-broadening audience. With a practical real-world emphasis, books benefit non-specialist engineers and students as well as EMC veterans. Books are written by recognized authorities in their field, with every manuscript thoroughly vetted by members of the EMC Editorial Board and additional qualified advisors. The publisher applies high quality production processes to each publication and sets prices for the budgets of working professionals and students.

Published titles:
Grudzinski/Trzaska – Electromagnetic Field Standards and Exposure Systems (2014)
Wyatt/Jost – The EMC Pocket Guide (2013)
Darney – Circuit Modeling for EMC (2013)
Bienkowski/Trzaska – Electromagnetic Measurements in the Near Field (2012)
Duff – Designing Electronic Systems for EMC (2012)

Forthcoming Titles
Andre/Wyatt – EMI Troubleshooting Cookbook for Product Designers
Wyatt/Jost – EMC Essential
Duff – Designing Wireless Systems for EMC

Electromagnetic Field Standards and Exposure Systems

Eugeniusz Grudzinski and Hubert Trzaska

SciTech
PUBLISHING
an imprint of the IET

Edison, NJ scitechpub.com

SciTech PUBLISHING
an imprint of the IET

Published by SciTech Publishing, an imprint of the IET.
www.scitechpub.com
www.theiet.org

Editor: Dudley R. Kay

10 9 8 7 6 5 4 3 2 1

ISBN 978-1-61353-177-8 (hardback)
ISBN 978-1-61353-181-5 (PDF)

Typeset in India by MPS Limited
Printed in the US by Integrated Books International
Printed in the UK by Berforts Information Press Ltd.

Contents

Preface   ix

Foreword   xi

Acknowledgments   xii

1 Introduction   **1**

2 EMF of an arbitrary structure   **7**
2.1 The near and the far field   7
2.2 EMF of a thin, symmetric, resonant dipole antenna   11
2.3 Homogeneity of EMF near a resonant dipole   13
2.4 EMF of a small loop antenna   17
2.5 Homogeneity of EMF near small loop   18

3 Methods of the standard EMF generation   **25**
3.1 Calibration of meters with dipole antennas   25
    3.1.1 Calibration with the standard field method   26
    3.1.2 Calibration with the substitution method   28
    3.1.3 Calibration of meters with whip antennas   29
3.2 Calibration of meters with loop antennas   32
    3.2.1 Calibration with the standard field method   33
    3.2.2 Calibration with the substitution method   35
    3.2.3 Calibration of meters with whip antennas   37
3.3 Calibration of meters with directional antennas   39
3.4 Meter calibration with the use of guided waves   41
    3.4.1 Calibration in the field of a plate capacitor   42
    3.4.2 EMF standards with a traveling wave line   43
    3.4.3 A waveguide as a standard EMF source   45
3.5 Secondary standards and exposure systems   46
    3.5.1 Chamber methods   47
    3.5.2 Examples of TEM chambers   49
    3.5.3 Other types of secondary standards   54

4 Accuracy analysis of EMF standards with dipole antennas   **63**
4.1 Accuracy analysis   63
4.2 Choice of the OATS   68
4.3 Measurement of the electrical parameters of the ground   72

4.4 Analysis of the accuracy of the SRA standard	74
4.4.1 Electromotive force measurement with a thermocouple	77
4.4.2 Voltage measurement using a selective voltmeter	79
4.4.3 Electromotive force measurement using a diode detector	80
4.5 Analysis of the accuracy of the STA method	81
4.6 Directional antenna calibration	85
4.6.1 E-field averaging along the antenna	88
4.6.2 Role of the radiation pattern	89

5 Accuracy analysis of EMF standards with loop antennas — **93**

5.1 Accuracy analysis of the SRA standard	93
5.1.1 Error in e_A measurement	94
5.1.2 Error in loop surface area measurement	94
5.1.3 Error in frequency measurements	95
5.1.4 Other factors limiting the standard's accuracy	95
5.2 Accuracy analysis of an STA standard	97
5.2.1 Accuracy of the current measurement	97
5.2.2 Accuracy of linear size measurement	97
5.2.3 Error due to noncentric placement of the antennas	98
5.2.4 Other factors limiting the standard's accuracy	99
5.3 An EMF standard with Helmholtz coils	104

6 Accuracy analysis of EMF standards with a segment of a transmission line — **111**

6.1 EMF in a strip line	112
6.2 Wave impedance of a TEM cell	113
6.3 Accuracy of the line accomplishment and its role in line accuracy limitation	116
6.4 Disturbances of the wave impedance	118
6.5 Line excitation measurement accuracy	123
6.5.1 Inaccuracy due to input voltage measurement	125
6.5.2 Inaccuracy due to simultaneous input and output voltage measurement	126
6.5.3 Inaccuracy due to incident voltage measurement	127
6.6 Accuracy of the EMF standard with a segment of a transmission line	130
6.7 Homogeneity of EMF in a TEM cell	131
6.8 An OUT in a TEM cell	135
6.8.1 An EMF probe in a TEM cell	135
6.8.2 A TEM cell as an exposure system	142
6.8.3 Absorption and polarization	147

7 Accuracy analysis of the standards with horn antennas — **151**

7.1 Accuracy of determination of the power gain	153
7.2 Accuracy of excitation measurement	159

7.2.1 Accuracy of power measurements 160
7.2.2 Accuracy of transmitting antenna excitation measurement 161
7.3 Accuracy of the standard estimation 162
7.3.1 Accuracy estimations of the SRA standard 162
7.3.2 Accuracy estimations of the STA standard 163
7.4 Nonstationary EMF standard 164

8 Comparative analysis of the EMF standards **169**
8.1 Double calibration method 169
8.2 Comparison of different type standards 171
8.3 International EMF standards comparison 173
8.4 Necessity of calibrations 178

9 Final comments **187**

References **193**
Index **197**

Preface

Electromagnetic field (EMF) measurements are among the least accurate of all measured phenomena. One of the most important factors limiting the accuracy of these measurements is the class of the EMF standards used. As the calibrated device cannot be more accurate than a standard applied to calibrate it, the need to improve the accuracy of the measurements requires an improvement of the EMF standards.

This book should prove helpful to those looking to improve measurement accuracy. We present a range of factors that play an essential role in the estimation of the accuracy of the standards. However, there are no two identical standards, and even the same standard is not always the same, even applied in the same place, by the same personnel, and using the same auxiliary equipment. These alterations are due to temporal, climatic, ageing, and variable local conditions. This leads to a requirement to repeat the accuracy analysis of a standard virtually every time it is applied. On the other hand, and for the same reasons, it is virtually impossible to propose a universal procedure or proscription for accuracy estimation. In practice, the user should take into account any possible factor affecting the standards work. We recognize that there is no one panacea and we do not promote excessive optimism. What is illustrated here is a pragmatic optimism that leads to good accuracy of a standard even beyond its estimated accuracy limits.

This book discusses the solution of standards (primary standards) that are designated for calibration of EMF meters and probes in wide frequency and measured value ranges. Here we propose two new approaches to standardization: 1) a whip antenna calibration using an H-field standard that makes the calibration easier and more accurate, and 2) a non-stationary EMF standard that allows the accuracy of non-stationary fields from the level of "qualitative measurements" to a level at which accuracy estimation is possible.

Similar, or identical, solutions are applied as secondary standards (exposure systems) in areas of electromagnetic compatibility (EMC), electromagnetic interference (EMI), biomedical studies, and many others. The main difference between objects tested in both cases is the size of the objects. Additional limitations on accuracy are experienced due to mutual interaction between these objects and the standard. As a result, accuracy in these studies is reduced in comparison to the primary standard. In many EMC and EMI tests the applied measuring procedures are nationally and/or internationally standardized. Due to this, there is no formal problem with accuracy. However, in basic research, and even in formalized standards-based measurements, errors and mistakes may take place that lead to significant differences between measurements presented in the literature, despite "identical" conditions in different labs.

Freedom in the selection and construction of an exposure system may be an advantage. However, an arbitrary approach to its use may lead to the differences previously mentioned. We have included a section in the book mainly addressed to investigators in the bioelectromagnetics area. Here we propose a new exposure system that makes it possible to take into account all radiation generated by a personal terminal, as well as an exposure system that allows limitation of absorption effects due to coupling between exposed objects and the exposure system. Another proposal is an exposure system with quasi-spheroidal polarization. The system permits the exposed objects to be kept *in vivo* to limit unwanted effects due to their immobilization. The system is useful in EMC testing as well, especially for testing large objects since it is easier to change spatial position of field vectors or the field polarization than turning the object itself.

EMF surveying, for labor safety or environment protection purposes, is usually done in difficult conditions, for example, an industrial environment, traceless access, any severe weather, etc. Moreover, the variety and variations of EMF levels close to sources of radiation may be a confounding variable to a measuring team. In order to make it possible to check a meter during such measurements, a variety of portable standards are proposed. We call them "the meters testers."

One of the possible ways to experimentally check the quality of a standard is to compare it with another standard. This may be done in any lab by performing calibration of a probe (antenna, meter) using different methods, for instance, through the calibration of a probe in a TEM cell and using a field method. More effective may be a comparison between two or more labs. This allows not only calibration using different sets but also a discussion on these issues.

This book is based upon the authors' personal ideas, their years of experience, and deep involvement in the area of EMF standards. Many presented solutions were proposed by them. Our aim is to help the readers to understand problems related to the accuracy limitations in EMF standards. However, we are not "Solomons" in the field. Our approaches lead to merely one perspective on many issues. The authors would like to express in advance their gratitude for any comments and suggestions in this field.

Eugeniusz Grudzinski
eugeniusz.grudzinski@pwr.wroc.pl

Hubert Trzaska
hubert.trzaska@pwr.wroc.pl

Foreword

Managing error budgets is important in all measurement activities. Within an EMC environment, however, this is especially vital for improving both accuracy and precision of any results. As upper frequencies extend ever further upwards, errors can become more obvious. Hence, developing a better understanding of the sources and management of errors is essential for all EMC engineers, from individual researchers to commercial test houses.

Electromagnetic Field Standards and Exposure Systems by professors Eugeniusz Grudzinski and Hubert Trzaska provides a detailed and accessible reference book to understand and manage measurement uncertainty in electromagnetic fields. The book concentrates on the analysis of the accuracy of standards using of dipole, loop, parallel plate transmission lines, and horn antennas. It also provides a range of methods for generating standard electromagnetic fields. This book provides a wealth of information to enable even a relatively modest laboratory to set up their own calibration bench.

It has been my pleasure and honour to work with Professor Trzaska on two books, and I have been fortunate to have spent some with him in his laboratory. Apart from being a fascinating person with a rich and compelling life story, his work is dedicated to understanding as much as possible about measurements, starting from first principles. This is clearly a philosophy that his co-workers and colleagues share. The book is a prime example of that dedication to understanding. Eugeniusz Grudzinski and Hubert Trzaska have produced a first rate reference text that should be on the bookshelves of every engineer who makes field measurements or uses the results of field measurements.

Alistair P. Duffy, PhD
Series Editor

Acknowledgments

This edition of the book has come a long, difficult way from first draft to publication. First of all it required precise "englishisation" of our Penglish (Polish-English) to a more widely understood language.

We would like to express our sincere thanks to the reviewers for their positive opinions and invaluable suggestions and comments. This edition would not be possible without the personal involvement of Mr. Dudley Kay and his help from the very beginning to this fruitful end. The hard work of Editor Dr Alistair Duffy, has led to our manuscript's lingual correctness and also to many corrections of a technical nature. We also thank Brent Beckley for his continuous assistance. Finally we must not forget introductory inspirations by doctors Helmut M. Altschuler, Ramon C. Baird and Paul F. Wacker (NIST Boulder, CO).

We would like to express our sincere thanks to everybody involved in making our nightmare fiction a reality.

Chapter 1
Introduction

Nowadays civilization may be characterized by the rapid growth of applications of the number and power of electromagnetic field (EMF) sources in, for instance, telecommunications, industry, science, medicine (ISM), and domestic uses. This growth has been accompanied by the necessity of measuring the electric field (**E**), magnetic field (**H**), and power density (**S**) in three main areas:

- Measurements of free propagating electromagnetic (EM) waves in telecommunications, radiolocation, radio navigation, radio astronomy, geophysics, and other areas. Here may include measurements relating to antennas (radiation pattern, gain, calibration).
- Electromagnetic interference (EMI) measurements to assure undisturbed coexistence of devices and systems in an EM environment that are subjects of electromagnetic compatibility (EMC). These measurements are related to the operation of the above-mentioned services as well as to many other sources of EMI, including electric motors and combustion engines, overhead power lines, information networks and devices, atmospheric discharges, and others.
- Biosphere exposure measurements relating to labor safety, especially in close proximity to EMF-generating devices and systems, as well as environmental protection and, first of all, general public protection.

There are in general use a variety of meters and sets for EMF measurement, starting from the lowest levels, close to or even below the noise level of the measuring devices, related to propagation or EMI measurements, through levels equivalent to the exposure limits permitted by appropriate protection standards, up to the highest levels that could cause even lethal effects or damage due to strong electromagnetic pulses (EMP) from such sources as, for instance, nuclear electromagnetic pulse (NEMP), generated during a nuclear explosion; lightning electromagnetic pulse (LEMP), generated by atmospheric discharges; and others. The latter are especially important in the study of the immunity of devices and networks exposed to the fields. The scale of this dynamic range is followed by the wide range of frequencies involved in intentional (telecommunication) or unintentional (ISM) radiation. The range starts at static E- and H-fields and ends at the highest frequencies generated and applied currently. Moreover, measurements may be selective (usually in telecommunications) through bandpass to wideband (as in quantifying unwanted exposure), in addition to such EMF parameters as polarization, modulation, time and spatial distribution and alternations, etc. Testing and calibration of antennas, probes, and meters designated for the above-illustrated purposes is the dominant purpose of the primary EMF standards (PS). The standards should ensure full acquaintance of the EMF parameters with required and known (estimated) accuracy. The PS assures the highest available accuracy.

Apart from calibration of meters, it is necessary to assure basic knowledge of the EMF parameters in such studies, as, for example:

- sensitivity of tissues and organisms to exposure in biomedical investigations,
- susceptibility testing of electric and electronic devices,
- research of properties of living and nonliving matter exposed to EMF,
- measurement of spontaneous combustion points or detonation of explosive materials and mixtures under EMF including distant blast materials and detonators,
- radiations examination of devices, living, and nonliving matter.

In the study of applied designs as in PS, however, apart from some specificity, usually the accuracy of the standards is remarkably reduced due to the presence in proximity to the standard EMF source a device or material under test. This class of standards we will call secondary standards (SS) or exposure systems (ES).

During outdoor EMF measurements, especially for environmental protection purposes, performed with portable EMF meters or indicators, under different atmospheric conditions (temperature, pressure, humidity, pollution), radiation conditions (many sources, often unknown) and transportation in local ways, or even traceless, it may be necessary to check a device in order to ensure that its indications are correct. For this purpose a special class of EMF standards is proposed, i.e., EMF meter testers (MT). We must point out here that very often in proximity to high-power devices (especially in telecommunications), EMF levels are very low due to application of a variety of protective means, while close to much lower power devices (e.g., portable terminals) the fields are stronger, which may lead to measuring team making errors, especially by those inexperienced in the specificity of EMF measurements, and it emphasizes the necessity to use a MT.

By definition, a measurement is a comparison of an unknown magnitude with a known one. A direct application of the definition may be applied only to basic physical magnitudes (length, weight). This possibility does not exist in the case of EMF. Standard EMF is usually generated by the use of a radiation system, and the intensity of the field at a given point is calculated on the ground of the properties of the system and a measurement of its excitation. Such a set, containing a radiator, a power source, excitation meter, and other auxiliary equipment, is a laboratory set, and its comparison with other similar sets in a direct way is impossible. For comparison purposes, special transfer standards (TS) are designed. These are usually simple probes that are measured in different labs, and the results of indirect measurements illustrate a concordance of standards when compared in this way.

Apart from several national and international recommendations regarding different EMF meter calibrations, a universal calibration method does not exist, and institutions usually adopt their own methods to suit their requirements, which may raise doubts even if similar procedures are applied in different centers.

Many doubts create an approach to the standards and standardization methods relating to the accuracy of estimates. The latter may be corrected by way of contacts with other centers involved and the standards comparison.

That was the way passed by the authors. Their primary involvement was related to EMF metrology in the near field. The best proof of theoretical considerations is an experimental verification of the estimation results. In our case it was necessary to dispose of an EMF with known parameters, i.e., a standard EMF. In this way the involvement of the authors in the area of EMF standards was a by-product. However, from the very beginning new ideas were worked out; for example, the world's first H-field standard at frequencies above 30 MHz was designed and completed and the correctness of the work was proven by way of participation in the international EMF standards comparison.

What does the electromagnetic field standard mean? Let's imagine a linearly polarized, transverse (TEM) EM wave propagating in direction z in a lossless, infinitely large space. Its E_x- and H_y-field components in Cartesian coordinates are given by:

$$E_x = E_0 \cos(\omega t + \varphi_1) \tag{1.1}$$

and:

$$H_y = H_0 \cos(\omega t + \varphi_2) \tag{1.2}$$

where:

E_0 – E-field amplitude,
H_0 – H-field amplitude,
ω – angular frequency,
φ_1 and φ_2 – phases.

A standard EMF is a field in which all these parameters are known as well as their spatial orientation (polarization), modulation, and spatial variations, and alternations in time are determined. Precise knowledge of the magnitude values, in the case of the standard EMF, is not enough. Information of essential importance is here an accuracy with which the values were found (measured, estimated), which allows estimation of the most important parameter of the standard: its accuracy, or the class of the standard.

Factors determining and limiting accuracy are a subject of detailed analysis performed for different types of standards. In the analyses it will be assumed that a generated standard EMF is a monochromatic EMF (although some comments related to achromatic fields will be given) of a known (with required accuracy) frequency, and its polarization is established on the ground of mutual positioning of EMF vectors of the EMF standard and type and parameters of an object under test (OUT). A coefficient of the polarization mismatch will be taken into account as one of the factors limiting the accuracy of the standard. All further considerations are focused upon identification and estimation of the role played by possible sources of error in the standard EMF generation.

A standard EMF generation, using an arbitrary method, requires indirectly establishing the value of the magnitude of concern, as a direct measurement of an EMF is still unknown. EMF strength is established on the ground of the measurement of the current, voltage, or value of another physical magnitude exciting the standard, the standard type, geometry of propagation, etc. An important factor limiting accuracy of the standard is a mutual coupling between the standard and an OUT and possible resulting changes of the object parameters. Another factor limiting accuracy is possible influence of other sources and fields upon the standard and OUT. In our consideration we will omit this factor and will assume that any external sources and objects do not affect considered procedures. It is evident, however, that in practice it is usually necessary to check the EM environment in which measurements are taken. For instance, antenna calibration in an open area test site (OATS) may be affected even by distant TV and FM broadcast stations.

The complex relation between measured value of the standard excitation measurement results and generated field parameters and the factors limiting accuracy of the standard means that EMF standards are one of the least accurate among standards of physical magnitudes. Presently, on the basis of data in the literature, we may estimate the typical error of EMF standards at the level of 5–10%, which is 2–3 times better compared to the situation only a few decades ago. The improvement in accuracy is a result of the availability of better excitation meters, better auxiliary equipment, and better experience, and the possibilities of more precise analysis of the factors limiting the accuracy of standards and, thus, the limiting or/and elimination of the sources of errors, as well as the ability to participate in standards comparison.

This book presents:

- Problems of EMF generation by the simplest EMF sources and the EMF homogeneity around them (Chapter 2).
- Current methods of standard EMF generation, with the use of quasi-stationary EMFs, freely propagating EM waves on an OATS, waves insulated from the external EM environment, e.g., in an anechoic chamber and sets based upon the guided waves concept (Chapter 3).
- Accuracy analysis of E-field standards with dipole antennas (Chapter 4); H-field standards with loop antennas (Chapter 5); EMF standards with guided waves (Chapter 6); and E-, H-, and S standards at microwave frequencies (Chapter 7).
- Chapter 8 presents results of standards comparisons and comments related to such actions.
- Parameters of the EMF standards designed by the authors (Chapter 9).

At the beginning it is necessary to note that there are no two identical EMF standards. They will not be identical even if they are based upon the same concept and are generated with identical devices from the same manufacturer. Thus, the presented considerations should be understood as examples that may be useful in any particular case of standard accuracy estimation. Even the same standard set, applied for different purposes, may have different levels of accuracy in different applications, and the accuracy of the standard should be estimated individually for each application. The latter is required if maximal accuracy is of concern. Of course, it is possible to assume that the accuracy is at a level for a range of the standard applications. However, this leads to the need to reduce requirements as to the class of the standard, usually reducing the class from a primary standard to a secondary one.

Chapter 2
EMF of an arbitrary structure

Generation of a standard EMF requires the application of standard transmitting and receiving antennas. Simple antennas are generally used in primary standards because their parameters can be calculated precisely. In this chapter, the parameters for simple antennas will be calculated theoretically, based on basic assumptions and simplifications. In particular, this chapter will consider a half-wave, symmetrical dipole antenna of large slenderness ratio and an electrically small loop antenna with uniform current distribution.

In the following analyses, attention is focused on the most important parameters and properties of the antennas and propagation phenomena that, through estimation of the EMF distribution around the antennas, play an essential role in estimations of standards accuracy.

2.1   The near and the far field

The EMF at an arbitrary point in space is determined by the Maxwell equations. Thus, a determination of the searched values of E- and H-field intensities is reduced to the equation solutions that fulfill assumed boundary conditions and conditions of radiation. The equations for a harmonic time-dependent field may be written in the form [39]:

$$\nabla \times \mathbf{E} = -j\omega\mu\mathbf{H} - {}^*\mathbf{J} \tag{2.1}$$

$$\nabla \times \mathbf{H} = j\omega\varepsilon'\mathbf{E} + \mathbf{J} \tag{2.2}$$

where:

   $\mathbf{E}$ and $\mathbf{H}$ – E- and H-field vectors,
   $\mathbf{J}$ and ${}^*\mathbf{J}$ – electric and magnetic current density,
   μ – permeability,
   ε' – complex permittivity: $\varepsilon' = \varepsilon - j\,(\sigma/\omega)$,
   ε – permittivity: $\varepsilon = \varepsilon_0\,\varepsilon_r$,
   ε_0 – permittivity of free space,
   ε_r – relative permittivity,
   σ – conductivity.

Note that the units are usually omitted as the SI unit system is uniformly applied.

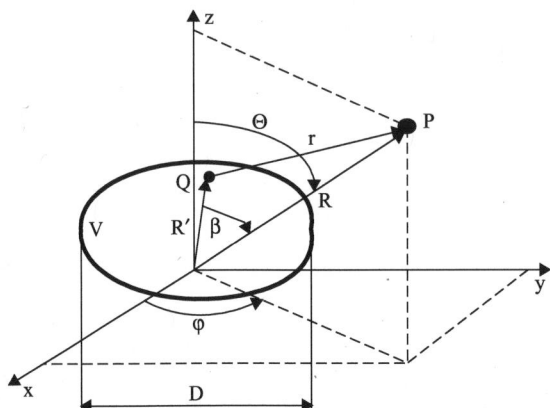

Figure 2.1   EMF in point P generated by a system of currents within the volume V

In a volume V (as shown in Figure 2.1), with a maximal linear size D, arbitrarily oriented electric and magnetic currents **J** and ***J**, respectively, flow. If the volume V is surrounded by an infinitely large, homogeneous, linear, lossless, and isotropic medium where the electric parameters are determined by permittivity ε and permeability μ (and the parameters remain unchanged at the surface bounding the volume V), we may find a solution to (2.1) and (2.2) and E- and H-field intensities at an arbitrary observation point P that are expressed in the form [31]:

$$\mathbf{E} = -j\omega\mu\mathbf{A} + \frac{1}{j\omega\varepsilon}grad\ div\mathbf{A} - rot^*\mathbf{A} \tag{2.3}$$

$$\mathbf{H} = -j\omega\varepsilon^*\mathbf{A} + \frac{1}{j\omega\mu}grad\ div^*\mathbf{A} + rot\ \mathbf{A} \tag{2.4}$$

where:
    A and ***A** – electric and magnetic potential vectors, respectively:

$$\mathbf{A} = \frac{1}{4\pi}\int_V \mathbf{J}\frac{e^{-jkr}}{r}dV \tag{2.5}$$

$$^*\mathbf{A} = \frac{1}{4\pi}\int_V {}^*\mathbf{J}\frac{e^{-jkr}}{r}dV \tag{2.6}$$

where:
    r – distance between a point of observation P (R, Θ, φ) and a point of integration Q (R', Θ', φ') that in vector notation is expressed by:

$$\mathbf{r} = \mathbf{R} - \mathbf{R}' \tag{2.7}$$

and the magnitude of this distance is:

$$r = \sqrt{R^2 + R'^2 - 2RR'\cos\beta} \tag{2.8}$$

where:

> R – distance between the point of observation P and the center of the coordi-
> nate system,
>
> R' – distance between the integration point Q and the center of the coordinate
> system,
>
> β – angle between R and R'.

For R' < R, the distance r may be expressed in the form of a series expansion:

$$r = R\left[1 - \frac{R'}{R}\cos\beta + \frac{R'^2}{2R^2}(1 - \cos^2\beta) + \frac{R'^3}{2R^3}(1 - \cos^2\beta)\cos\beta - \cdots\cdots\right] \quad (2.9)$$

If we calculate the difference between the real value of r given by (2.8), and its approximate value given by the two first terms of the series expansion, as given by (2.9), and then multiply it by the propagation constant $k = \omega\sqrt{\varepsilon\mu} = \omega\sqrt{\varepsilon_0\mu_0}$ (where: μ_0 permeability of free space), we will have a phase error $\Delta\Psi$ in the kernels of integrals (2.3) and (2.4) in free space. This error defines a range in which the approximation $R \gg D$ may be applied. The error may be expressed in the form:

$$\Delta\Psi = k\left(\frac{R'^2}{2R}\sin^2\beta + \frac{R'^3}{2R^2}\cos\beta\sin^2\beta\right) \quad (2.10)$$

Now, if we accept that the maximal value of the distance R' between the center of the coordinate system and the point of integration is equal to half of the maximal cross section of the volume (i.e., $R'_{max} = D/2$), we will obtain the maximal value of the phase error $\Delta\Psi_{max}$, which is given in (2.11):

$$\Delta\Psi_{max} = \frac{kD^2}{8R} = \frac{2\pi}{N} \quad (2.11)$$

where:

> N – a number reflecting acceptable uncertainty of the phase front in the point
> of observation; usually we assume $N \geq 16$.

Substituting N = 16 into (2.11), after some transformation we will have (2.12), which indicates the widely accepted definition of the boundary of the far field:

$$R \geq \frac{2D^2}{\lambda} \quad (2.12)$$

Equivalent to the assumed N = 16 is the phase difference $\Delta\Psi = 22.5°$ in the plane of observation. In the case of a dipole calibration the difference is illustrated in Figure 2.2. The definition requires some comments:

1. As can be seen above, the definition was introduced on the grounds of the arbitrary assumption N = 16. However, we must remember that the definition does not express any information regarding the parameters of the EMF generated by a source and plays only an informative role that should warn where the field of concern may have more complex structure and will require a more

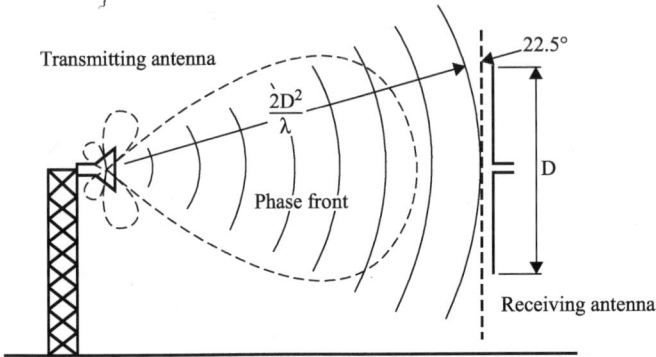

Figure 2.2   A curvature of the phase front in the case of a dipole antenna calibration

exact approach to its measurement or estimation of its parameters. The main advantage of the definition is its simplicity and the fact that it can be used in relation to any source without needing to undertake a full analysis other than determining the geometrical sizes of the source or an OUT (D – indicates the sizes of the larger of them, as in Figure 2.2).

2. It is possible to propose many more definitions of the far-field boundary. One of them assumes E- or H-field strength difference at the level of 5%, while strict formulas of the field calculation are compared with simplified ones [35]. Contrary to the previous definition, where the boundary is a sphere around the radiation source, here the boundary may have any spatial shape. Although this definition, and similar ones, may be considered as more accurate, they require separate estimations for any source and any configuration considered.

3. Definition 1 suggests that if (2.12) is fulfilled, we have something like a TEM wave in a point of observation. We may ask what does exist if $R \gg 2D^2/\lambda$, in the case of an assumed infinitely large, undisturbed space. It will be a spherical wave that in a point of observation may be accepted as a TEM wave. However, under real conditions, while multipath propagation takes place, even at a distance we may have a strongly interfered field, where direct rays may not exist at all and the structure of the field may be much more complex compared to a neighborhood of simple antennas (Figure 2.3). Moreover, due to the Doppler effect, while a reflection from a moving object or/and FM modulation, the polarization of the EMF at point A may be quasi-ellipsoidal; i.e., the considered field has three spatial components. This has led the authors to the most rigorous definition of the near-field boundary: "*The near field is everywhere I perform measurements.*" The definition relates to EMF measurements as well as to standardization and calibration procedures. It suggests necessary caution, especially when measurements are performed outdoors or on an OATS.

The above discussion was included only in order to show readers the meaning of the far or the near field. In later considerations we will come back to this expression only for illustrative purposes. Any calculations will be done with the use of complete formulas, with no reliance on the above accepted simplifications.

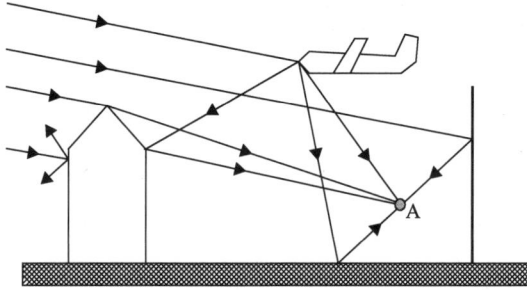

Figure 2.3   An example of interfered EMF due to multipath propagation

2.2   EMF of a thin, symmetric, resonant dipole antenna

The thin, symmetric, resonant dipole antenna is often applied as a standard antenna. Such an antenna may be used both as the standard transmitting antenna (STA) or the standard receiving antenna (SRA) in calibrations performed at an OATS or within an anechoic chamber. Because of the size of the antenna, its use is limited to frequencies above 30 MHz in practice.

Consider the EMF generated around the antenna in an infinitely large, homogeneous, and lossless space when the antenna is fed from a sine wave source. The geometry of the antenna is shown in Figure 2.4.

In our considerations we will refer to the previous chapter and (2.3) in order to calculate the E-field around the antenna. In this case, the magnetic current density *J may be assumed to be zero, and only the electric current **J** should be taken into account. The E-field given in (2.3) can be expressed in the form of (2.13):

$$E = -j\omega\mu A + \frac{1}{j\omega\varepsilon}\, \text{grad div } A \tag{2.13}$$

As may be seen from Figure 2.4, we can assume that the current in the antenna has only one spatial component, i.e., a z component, due to the antenna being thin. Hence, the electric potential vector **A** has one component as well:

$$A_z = \frac{1}{4\pi}\int_{-h}^{+h} I_z\left(\frac{e^{-jkR_1}}{R_1} + \frac{e^{-jkR_2}}{R_2}\right) dz \tag{2.14}$$

where:

R_1 and R_2 are radii shown in Figure 2.4.

Substituting (2.14) into (2.13), and after some transformations, the E_z component takes the form:

$$E_z = \frac{1}{j\omega\varepsilon}\left(k^2 A_z + \frac{d^2 A_z}{dz^2}\right) \tag{2.15}$$

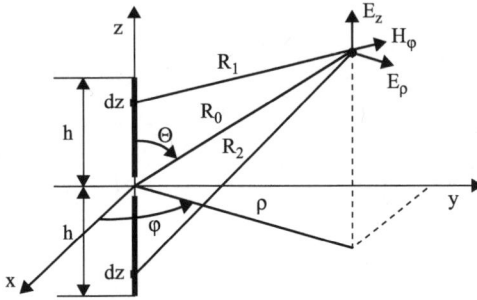

Figure 2.4   Thin symmetric dipole in coordinate system

The considerations presented here are very general in character and present the EMF at an arbitrary distance from the dipole. The geometry of the structure has already been presented. The only unknown magnitudes are the current exciting the antenna and its distribution along the antenna. Now we will make use of the above-mentioned slenderness ratio of the antenna. In a thin antenna, one may assume a sinusoidal current distribution of I_z, i.e.:

$$I_z = I_0 \frac{\sin k(h - z)}{\sin kh} \qquad (2.16)$$

where:

I_0 – current exciting the antenna at its input.

The assumption of a sinusoidal current distribution in a thin antenna (i.e., an antenna fulfilling the condition $a \ll h$, where: a – diameter of the conductor the antenna is made of) is acceptable if we assume a boundary condition $I_z = 0$ for $z = h$. For thin antennas and for antennas longer than a half wavelength, this condition is not always fulfilled, and use of this assumption may lead to errors, especially when the input impedance of the antenna is calculated. From the point of view of the radiation pattern calculation, the condition is of secondary importance and may be omitted, especially when antennas of $h \leq \lambda/4$ are considered. In the role of standard antennas, usually the latter condition is fulfilled, which enables the assumption of the sinusoidal current distribution to be used.

After performing the integration in (2.14) and then substituting into (2.15), we have the E_z field component of the dipole antenna:

$$E_z = -\frac{jZI_0}{4\pi \sin kh} \left[\frac{e^{-jkR_1}}{R_1} + \frac{e^{-jkR_2}}{R_2} - 2\cos kh \frac{e^{-jkR_0}}{R_0} \right] \qquad (2.17)$$

This is valid for antennas of any length (taking into account the above discussion). As has already been mentioned, in the role of standard antennas usually half-wave dipoles are used; thus, for $h = \lambda/4$, (2.15) takes the form:

$$E_z = -\frac{jZI_0}{4\pi \sin kh} \left[\frac{e^{-jkR_1}}{R_1} + \frac{e^{-jkR_2}}{R_2} \right] \qquad (2.18)$$

where:

Z – intrinsic impedance of free space: $Z = 120\pi$ [Ω], and:

$$R_1 = \sqrt{\rho^2 + (z - h)^2}$$
$$R_2 = \sqrt{\rho^2 + (z + h)^2} \tag{2.19}$$
$$R_0 = \sqrt{\rho^2 + z^2}$$

Usually, in cases of practical importance, while the EMF strength is calculated at a distance from a STA, one may assume that $R_1 = R_2 = R_0 = \rho$ if $\rho \geq 10$ z. Or, in the terms of the antenna length, for $h = \lambda/4$, i.e., for $\rho \geq 2.5\lambda$, then (2.18) takes the form:

$$E_z = -\frac{jZI_0}{2\pi \sin kh} \left[\frac{e^{-jkR_0}}{R_0} \right] \tag{2.20}$$

Equation (2.20) allows calculation of the E_z component at a point located on the xy plane at a distance ρ from the coordinate system center. However, it may be assumed, with acceptable accuracy, that the component is identical (constant) along a receiving antenna, perpendicular to the plane. Usually the arm length 1 of the latter antenna fulfills the condition $0 \leq 1 \leq \lambda/4$. In the above discussion, only the transverse E-field component (E_z) was taken into account. Both the radial E-field component and the H-field components were omitted, as they are not essential when the standard EMF is considered in this way.

2.3 Homogeneity of EMF near a resonant dipole

As previously mentioned, (2.20) gives the E-field intensity in the xy plane when used to estimate fields around an antenna under test. Now we will prove the assumption of equivalence of the field in the plane and along the antenna under test. Consider a transmitting, symmetric, half-wave dipole antenna A_t ($h = \lambda/4$), placed parallel to a receiving, symmetric dipole antenna A_R of arm length $1 \leq \lambda/4$. The centers of both antennas are placed on a common plane with separation ρ, as shown in Figure 2.5.

In order to investigate the maximal difference between the E-field in the center of the receiving antenna and that at its end, we will define divergence δ_1 in order to

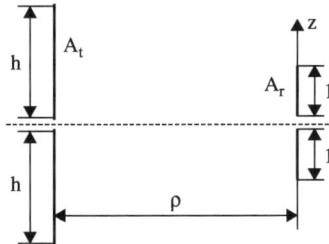

Figure 2.5   Two parallel dipoles

quantify the difference:

$$\delta_1 = \frac{E_z(z)}{E_z(0)} - 1 \tag{2.21}$$

where:

$E_z(0)$ – E field in the center of antenna A_r, calculated on the basis of (2.20),
$E_z(z)$ – E-field at the end of the antenna ($z = 1$), calculated on the basis of (2.18).

Calculations were completed with an assumption that mutual couplings between the antennas are negligible (this assumption is very rough, especially if $1 \cong \lambda/4$ and $\rho < \lambda/2$) for two lengths of antenna A_r, i.e., $1 = 0.25\lambda$ and $1 = 0.1\lambda$. The first case is evident; the latter reflects a practical situation in frequency range close to 30 MHz, where use of resonant antennas is limited due to their sizes and problems with the mechanical stability of such a device—thus, in practice, below 80–100 MHz applied antennas are usually resonant within this frequency range, and for measurements below it as well. Results of calculations are shown in Figure 2.6 as a function of ρ/λ.

The curves shown in Figure 2.6 make it possible to draw the conclusion that the difference δ_1 is significant for distances less than a wavelength. Further, this difference vanishes for distances over several wavelengths. The calculations were performed to show the E-field distribution around, and at a distance from, the transmitting antenna; they are not substantially influenced by the inherent assumptions. It has been shown that a point value of the E-field along antenna A_r differs considerably in relation to that at the center, especially when the antenna is near a source. This leads to well-known rule that in the near field, for point-value EMF measurements, short antennas should be used. The curves in Figure 2.6 illustrate problems with the generation of uniform fields: The distance between the two antennas should be of several wavelengths. This can create difficulties because of the space available (especially in environments such as an anechoic chamber).

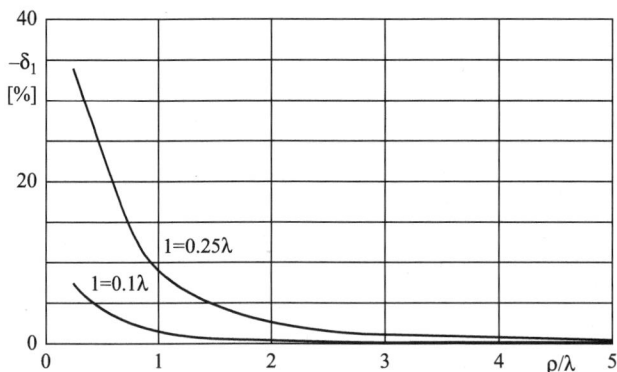

Figure 2.6   Divergence between the E-field $[E_z(l)]$ at the end of a receiving antenna in relation to $[E_z(0)]$ at its center

Also, the necessary excitation power can give rise to problems due to interference generation when measurements at an OATS are performed. However, our objective is not the measurements but calibration. Now, we will consider what value of the field is "seen" by the receiving antenna due to averaging effects.

In order to find the average value E_{AV} of E-field at the length of the receiving antenna, we will use the equation as before:

$$E_{AV} = \frac{1}{2l} \int_{-1}^{+1} E_z dz \qquad (2.22)$$

We will refer the average value (E_{AV}) calculated using (2.18) and (2.22) to the E-field value, $E_z(0)$, at the center of the antenna given by (2.20), and the result we will call the "error of calibration" due to the field averaging δ_2:

$$\delta_2 = \frac{E_{AV}}{E_z(0)} - 1 \qquad (2.23)$$

These calculations are again performed for two receiving antennas of length 0.25λ and 0.1λ, and the results are shown in Figure 2.7. The results of the calculations presented in Figure 2.7 may be interpreted in two ways:

1. Obtaining the value of the calibration error due to averaging effect, when an antenna of length l is calibrated with the use of the standard transmitting antenna (STA) method.
2. Determining the value of the error when, in the field generated by a transmitting antenna, the standard receiving antenna (SRA) calibration method is applied.

Both methods will be discussed in detail in section 3.1. Now we will return to (2.20). During its derivation we assumed that the simplifications are acceptable for $\rho \geq 2.5\lambda$. Evidence in support of this assumption can be seen from Figure 2.7.

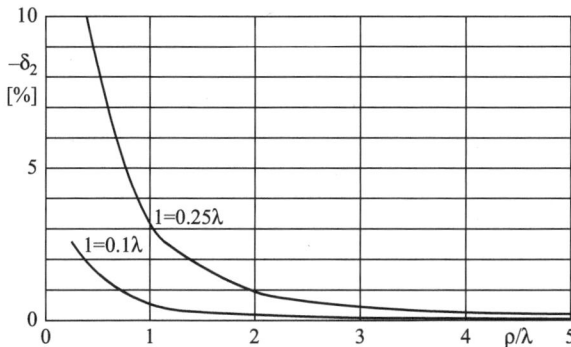

Figure 2.7 Averaging error δ_2 as a function of ρ/λ

As has already been said, the results of the averaging error estimations are valid for any length of a receiving antenna while the STA method is in use. However, while the SRA method is in use, the standard antenna length is resonant and a calibrated antenna may be of any length. This is especially important at frequencies below 80–100 MHz. Even from Figure 2.7 it can be seen that the averaging error is a function of the antenna length. Now we should consider what is the value of averaging calibration error in the substitution method (SRA method) δ_3, when in the place of SRA of arm length 0.25λ, a calibrated antenna of arm length 0.1λ at 30 MHz is situated. The error is defined as:

$$\delta_3 = \frac{E_{\lambda/4}}{E_{0.1\lambda}} - 1 \qquad (2.24)$$

where:

$E_{\lambda/4}$ – the field averaged by half-wave SRA,
$E_{0.1\lambda}$ – the field averaged by a calibrated antenna of length $l = 0.1\lambda$.

Both these values were calculated using (2.22). The results of the calculations are presented in Figure 2.8. Three comments may be formulated here:

1. The curve presented in Figure 2.8 is a majorant of errors. As calculations were performed for $l = 0.1\lambda$, and the antenna will be electrically longer for higher frequencies, the error vanishes at approximately 80 MHz. However, this presents a case of practical importance where a resonant antenna is applied for EMF measurements at a lower frequency (greater wavelength), as, for example, an 80-MHz antenna being used at 30 MHz.
2. Comparing errors presented in Figures 2.6, 2.7, and 2.8, it may be seen that the STA method requires less space, or is more accurate, when shorter antennas are calibrated.
3. A comparison of curves shown in Figures 2.7 and 2.8 would suggest the SRA method for resonant antenna calibration in cases where space or excitation power is limited. For $\rho > 3\lambda$, the two methods are comparable as regards the errors caused by the inhomogeneous EMF distribution.

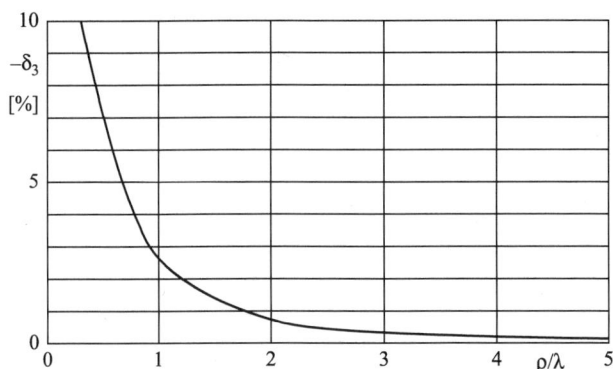

Figure 2.8   Averaging calibration error δ_3 in SRA method

2.4   EMF of a small loop antenna

It may be observed from the Maxwell equations that the E-field is always accompanied by the H-field (and vice versa). Thus, it would be possible to take into account considerations relating to H-field generation and thus broaden the discussion presented in the previous section to this field component. The approach would be fully correct from a theoretical point of view, but its practical importance would be very limited. Thus, for H-field generation, loop antennas are generally used. Although a dipole antenna has an H-component as well, the dominant field is the E-field component, and the H-field would need more power for excitation as compared to the equivalent loop. Of course, an OUT would also be affected by a strong E-field. Moreover, resonant dipole antennas are preferred in the frequency range of 30–1000 MHz. At lower frequencies, where the antennas must be electrically short for mechanical reasons, their radiation is much less effective, whereas H-field sources should work from static fields up to, at least, 30 MHz.

H-field components, from any source, may be derived based on (2.4). We will assume a current I flowing in a circular loop antenna. The antenna is electrically small, which allows us to assume that the current distribution along the antenna is of uniform amplitude and constant phase. In order to simplify our considerations, we will assume that the diameter of the conductor is much smaller than the antenna diameter, The antenna is placed in the xy plane of the Cartesian coordinate system, as shown in Figure 2.9.

As in the case of dipole antenna, one may assume that the magnetic current density *J is negligible in relation to the electric current density **J**. This leads to the magnetic potential *A given by (2.6) vanishing, resulting in having only the electric

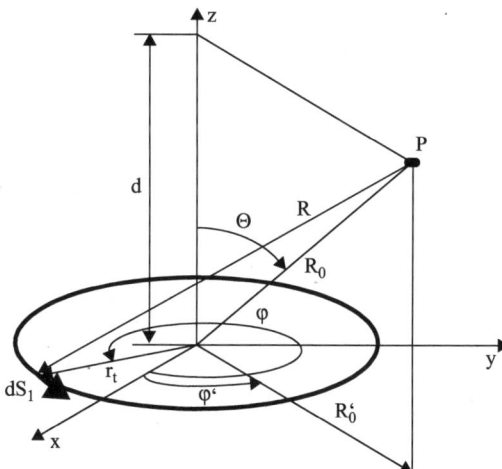

Figure 2.9   A small, circular loop antenna in coordinate system

potential **A** given by (2.5). The latter, for assumed quasi-stationary conditions, is given by:

$$A = \frac{1}{4\pi} \int_{S_t} J \frac{e^{-jkR}}{R} dS_t \tag{2.25}$$

where:
  R – leading radius:

$$R = \sqrt{r_t^2 + R_0^2 - 2r_t R_0 \sin\Theta \cos(\phi - \phi')} \tag{2.26}$$

where:
  r_t – radius of the transmitting antenna,
  S_t – surface area of the antenna, $S_t = \pi r_t^2$.

Then, the H-field at an observation point P is given by:

$$\mathbf{H} = \mathrm{rot}\,\mathbf{A} \tag{2.27}$$

In the simplest, stationary case, for $r_t \ll R_0$, the H-field around the loop is given by the radial component H_r:

$$H_r = \frac{I \cdot S_t}{2\pi \cdot R_0^3} \sqrt{1 + k^2 R_0^2} \cos\Theta \tag{2.28}$$

and the transverse component H_Θ:

$$H_\Theta = \frac{I \cdot S_t}{4\pi \cdot R_0^3} \sqrt{1 - k^2 R_0^2 + k^4 R_0^4} \sin\Theta \tag{2.29}$$

2.5   Homogeneity of EMF near small loop

In the case of the previously considered dipole antennas, the transmitting and receiving antennas can be thought of as having only one spatial dimension. Thus, in considering field averaging by the receiving antenna, we used (2.22). In the case of H-field generation (and measurement), we use, for both transmission and reception, a loop antenna of two spatial dimensions in order to simplify the analysis. However, the third dimension may be of some importance when multiturn antennas, or fields at high frequencies, are to be considered.

A set of two coaxial loops is shown in Figure 2.10. A receiving loop antenna of radius r_r is placed coaxially with a transmitting loop of radius r_t at distance d. As may be supposed, at the surface of the receiving antenna (a calibrated antenna or SRA), the field nonuniformity should be taken into account. Here, averaging is not expressed as simply as previously and, in the literature, different approaches may be found to the field calculations. This section will follow the approach

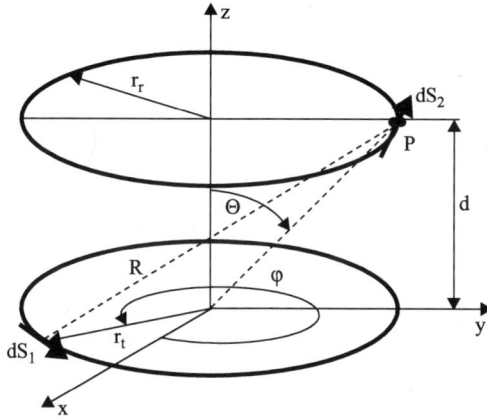

Figure 2.10 Two coaxial loops in coordinate system

of Green [18]. The H-field averaged at the surface of a receiving loop H_{AV} is given by:

$$\mathbf{H}_{AV} = \frac{1}{S_r} \int\limits_{S_r} rot \ \mathbf{A} dS_r \tag{2.30}$$

where:

S_r – surface area of the receiving loop,

$\mathbf{A}$ – electric vector potential given by (2.25),

$\mathbf{R}$ – distance between dS_t at the circumference of the transmitting antenna and dS_R at the receiving one, as in Figure 2.10; if in (2.26) for $\varphi' = 0$ we substitute $r_r = R_0 \sin\Theta$ and $R_0^2 = r_r^2 + d^2$, we will have:

$$R = \sqrt{d^2 + r_t^2 + r_r^2 - 2r_t r_r \cos \phi} \tag{2.31}$$

where:

r_r – radius of the receiving antenna,

r_t – radius of the transmitting antenna.

For assumed quasi-stationary conditions and two coaxially placed loops, substituting (2.25) into (2.27) and then into (2.30) results in:

$$H_{AV} = \frac{I r_t}{\pi r_r} \int\limits_0^\pi \frac{e^{-jkR}}{R} \cos \phi \ d\phi \tag{2.32}$$

It is impossible to calculate the integral of (2.32) analytically. Thus, the kernel of the integral may be expanded in a series. If the kernel is expressed with the use of the Henkel's spherical functions we will have [17]:

$$H_{AV} = -j \frac{I k r_t}{\pi r_r} \int\limits_0^\pi \frac{1}{m!} \left(k \frac{r_t r_r}{R_0} \right)^m h_m^{(2)} \ k \ R_0 \cos^{m+1} \phi \ d\phi \tag{2.33}$$

where: $h_m^{(2)}$ – mth kind and second-order Henkel's function, and:

$$R_0 = \sqrt{d^2 + r_t^2 + r_r^2}$$

If $R_0 > \sqrt{2 r_t r_r}$, then exchanging integration into summation, we may rewrite (2.33) in the form:

$$H_{AV} = -j \frac{I k r_t}{\pi r_r} \sum_{m=0}^{\infty} \frac{1}{(2m+1)!} \frac{1 \cdot 3 \cdot 5 \cdots (2m+1)}{2 \cdot 4 \cdot 6 \cdots (2m+2)} \left(\frac{k r_t r_r}{R_0}\right)^{2m+1} h_{2m+1}^{(2)}(kR_0)$$

(2.34)

Equation (2.34) may be presented in the form of a power series:

$$H_{AV} = \frac{I S_t}{2\pi R_0^3} \left[(1 + jkR_0) + \frac{15}{8} \left(\frac{r_t r_r}{R_0^2}\right) \right.$$

$$\left(1 + jkR_0 - \frac{6}{15} k^2 R_0^2 - j\frac{1}{15} k^3 R_0^3 \right) + \frac{315}{64} \left(\frac{r_t r_r}{R_0^2}\right)^4$$

$$\left. \left(1 + jkR_0 - \frac{420}{945} k^2 R_0^2 - j\frac{105}{945} k^3 R_0^3 + \frac{15}{945} k^4 R_0^4 + j\frac{1}{945} k^5 R_0^5 \right) + \cdots \right]$$

(2.35)

where: S_t – surface area of the transmitting antenna.

If $kR_0 \leq 1$ and $r_t r_r / R_0^2 \leq 1/16$, then (2.35), with an accuracy of 0.2%, may be written in the form:

$$H_{AV} = \frac{I S_t}{2\pi R_0^3} \left[1 + \frac{15}{8} \left(\frac{r_t r_r}{R_0^2}\right)^2 + \frac{315}{64} \left(\frac{r_t r_r}{R_0^2}\right)^4 + \cdots \right] \sqrt{1 + k^2 R_0^2}$$

(2.36)

If $d/r_r > 4$ and $d/r_t > 4$, the latter, with accuracy to 1%, may be expressed as follows:

$$H_{AV} = \frac{I S_t}{2\pi R_0^3} \sqrt{1 + k^2 R_0^2}$$

(2.37)

Equation (2.37) is usually applied when the averaged field at the surface of a receiving antenna is calculated on the basis of the parameters of a transmitting antenna, its excitation, and the geometry of propagation. Comparing (2.28) and (2.37), it may be seen that they are equivalent on the z-axis.

As to the case of dipole antennas, we will present several examples of inhomogeneity of the H-field distribution around a circular loop antenna and its role in H-field standard accuracy estimations.

Imagine two coaxial, circular loop antennas, placed in the Cartesian coordinate system shown in Figure 2.10. A transmitting antenna of radius r_t is placed directly

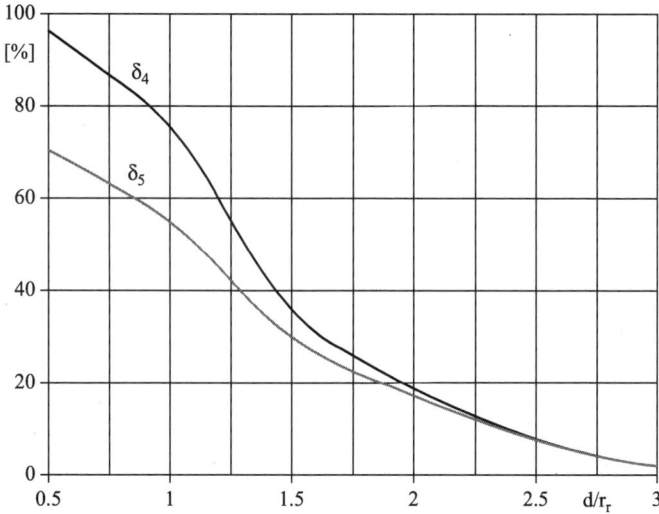

Figure 2.11   Divergence of the H-field in the plane of a receiving loop

in the xy plane. The receiving antenna, of radius r_r, is placed parallel to the former, at distance d. Centers of both antennas are on the z-axis.

In order to illustrate the level of the inhomogeneity of the H-field generated by the transmitting antenna, considered in the plane of the receiving antenna, Figure 2.11 presents two curves. Curve δ_4 illustrates a divergence of the H-field at the circumference of a receiving antenna in relation to that at its center, and δ_5 presents the divergence of the averaged H-field in relation to that at the antennas center. They are defined as follows:

$$\delta_4 = \frac{H_{rr}}{H_z} - 1 \tag{2.38}$$

and

$$\delta_5 = \frac{H_{AV}}{H_z} - 1 \tag{2.39}$$

where:

H_{rr} – H-field at circumference of a receiving loop, at distance r_r from the axis z,
H_z – H-field at axis z, at distance d from the transmitting antenna,
H_{AV} – averaged H-field at the surface of the receiving loop.

In the calculations above, it was assumed that $r_t = r_r$; i.e., the two antennas are identical. The estimations have no practical importance, except for the case assumed. Similar calculations should be repeated for any given set of antennas applied in metrological practice. However, they illustrate well that H_{AV} majorizes the H-field at the circumference of the receiving antenna and that both fields are

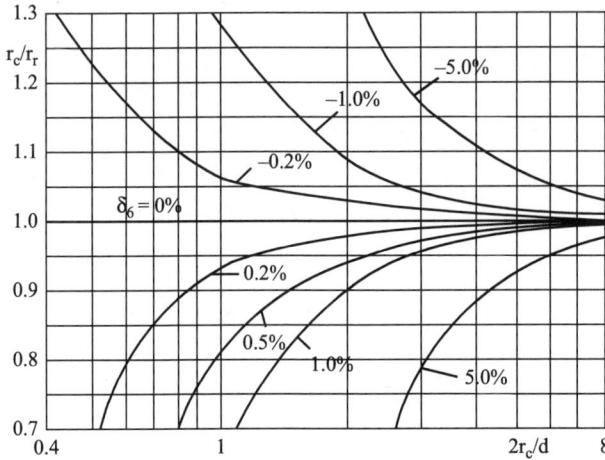

Figure 2.12   Substitution error δ_6 in the SRA method when SRA and an antenna under test are of different sizes

less than the magnetic field at the z-axis. An evident conclusion may be drawn: Larger receiving antennas give a larger H-field inhomogeneity at their surface, and antennas located further from each other give rise to a more homogeneous H-field at the surface of the receiving antenna. Moreover, the estimations only illustrate the H-field inhomogeneity, and the two methods (i.e., STA and SRA) are equivalent because, in both cases, H_{AV} is taken into account. The phenomenon is of primary importance when a transmitting loop is applied for an exposure of an OUT.

Now we will consider an error in the SRA method: substitution error δ_6. This is when an H-field is generated by a transmitting antenna and the field is measured by a standard receiving antenna that is then substituted by an antenna under test. The error we will define in the form:

$$\delta_6 = \frac{H_{AVc}}{H_{AVSRA}} - 1 \tag{2.40}$$

where:

H_{AVc} – H-field averaged at the surface of a calibrated antenna,

H_{AVSRA} – H-field averaged at the surface of the standard receiving antenna.

Estimations of the substitution error δ_6 for ratios $r_c/r_r = 0.7 \div 1.3$, as a function of $2r_c/d$, are plotted in Figure 2.12. In the figure, values of δ_6 at the level $0 \div \pm 5\%$ are indicated. The considerations presented here allow following conclusions to be drawn:

1. Effects caused by inhomogeneous field distributions around a source may be neglected when the distance between a source and a plane of calibration is large. The condition may almost always be fulfilled when antennas and meters

designated for measurements in the far field are calibrated. In this case, a low-level H-field intensity is enough. However, when meters and probes for the near field are tested, high intensities of H-field are usually required. Hence, a problem with power exciting a transmitting antenna may be noticed. In the latter case, the distance between the transmitting antenna and a device under test must usually be as small as possible, and then the analysis presented here is of primary importance.

2. In order to improve calibration accuracy, when the distance d is small, it is necessary to use more accurate formulas than (2.37).

3. As may be seen from Figure 2.12, in the case of the substitution method, when the SRA and an antenna under test are of different diameters, the substitution error should be taken into consideration. It may be observed that the error does not exist when the STA method is applied instead.

4. In any case considered, the degrading accuracy role of inhomogeneity may be reduced with the use of more accurate formulas.

5. The calculations presented are only of demonstrative character. Any case requires an individual approach taking into account the method of calibration, the parameters of the antennas applied, and the geometry of the system.

6. The phenomena discussed are of primary importance when a dipole or loop antenna is applied as a radiator in an exposure system, especially when an OUT is placed in proximity to the radiator.

Chapter 3

Methods of the standard EMF generation

The previous chapter referred to two calibration methods: the standard transmitting antenna method (STA) or standard field method, and standard receiving antenna method (SRA) or substitution method. Let's characterize them briefly.

1. The STA method is based upon EMF generation with the use of a transmitting antenna whose parameters (radiation pattern, efficiency) are known to a required accuracy and established either theoretically or experimentally. The EMF at an observation point is calculated taking into account the parameters of the antenna, its excitation, and the geometry of propagation.
2. The SRA method requires a receiving antenna of well-known parameters and a system that makes it possible to measure current in the antenna or the voltage at its input. The SRA is placed in an EMF generated by an arbitrary source and then replaced by an antenna (meter, probe, device) under test. Two assumptions are made in this procedure:

   • the EMF is stable enough to not change during the replacement,
   • the antenna under test is immersed in the same field as the SRA and makes identical EMF deformations to the SRA.

In the previous discussions we took into account EMF standards with dipole or loop antennas. However, an almost identical approach may be adopted when other types of standards are in use. For instance, in guided wave standards, the EMF is established on the ground of excitation measurement of a system of known parameters, which is similar to the standard field method. In the case of different types of chambers the substitution method is more appropriate.

3.1 Calibration of meters with dipole antennas

Meters with dipole antennas are primarily designed for E-field measurements in two different areas of application. Resonant-sized antennas are applied mainly within the frequency range 30–1000 MHz for propagation studies, EMI measurements, and similar applications where usually low-level E-fields are measured. Small-sized (electrically short) dipole antennas are applied for indoor measurements and in near-field metrology in such applications as quantifying nonionizing radiation exposure for labor safety and environmental monitoring purposes, near-field measurement of the radiation pattern of large antennas, and similar applications. Short dipoles are applied over a much wider frequency range compared to resonant antennas—in practice, from static fields to gigahertz frequencies. At microwave frequencies, they are often used in the power density (S) measurement.

At frequencies above 300 MHz, resonant antennas may be applied in different combinations of wide bandwidth and directional radiation pattern—for instance, thick-wideband dipoles, multielement arrays of active and passive dipoles, Yagi-Uda antennas, log-periodic antennas, etc. For their calibration the same methodology as for single dipole antennas may be applied.

The calibration procedure prefers such a situation when a transmitting antenna is at such a distance from the receiving antenna that the received EMF has a homogeneous structure, possibly approaching the plane wave, as was possible to conclude from the discussion in the previous chapter. The requirement is most difficult to fulfill at the lowest frequencies, i.e., around 30 MHz. In order to ensure that the calibration procedure is not affected by external EMI and to allow uninterrupted work, irrespective of weather conditions, anechoic chambers are preferred to open area test sites (OATS). Again, where lower operating frequencies are required, larger sizes of the chambers are needed. For instance, a chamber used by the authors of dimensions $4.9 \times 4.7 \times 2.2$ m has the first resonant frequency at 36 MHz, and its applicability for calibration procedures may be assumed as above 100 MHz. In order to save expenses on the construction of large chambers, the use of natural caves for the purpose was proposed by the National Bureau of Standards (currently the National Institute of Standards and Technology) some time ago [1].

3.1.1   Calibration with the standard field method

The standard field method application, or STA as it was previously named, for calibration meters with dipole antennas is shown in Figure 3.1. A standard transmitting antenna A_t is placed at a distance d from the antenna A_c of a calibrated

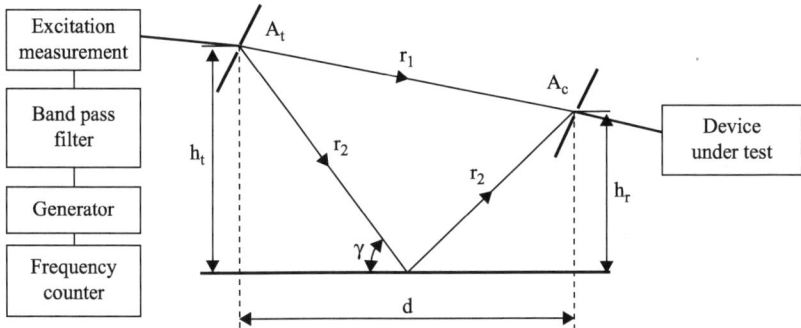

Figure 3.1   Calibration of a meter with dipole antenna on an OATS using the STA method

meter. The antenna A_c is placed at a height h_c above the ground, while the transmitting antenna is at a height h_t. The antennas are placed horizontally above the ground (horizontal polarization), parallel to one another, and their centers are in a common plane. The heights of both antennas are important when the antennas are placed above a plane of nonzero reflection factor, e.g., at an OATS. The heights may be neglected when the calibration is performed in an anechoic chamber. In the latter case, the presence of reflected waves from the sidewalls of the chamber (and their role) must be checked before calibration starts. Sometimes an artificial reflection and two ray propagation (as shown in Figure 3.1) may be necessary, in which case, the bottom (floor) of the chamber is to be covered by a conducting material.

The wave propagation between the transmitting antenna and the receiving one, in the case as shown in Figure 3.1, is by two paths:

- a direct ray of r_1 length,
- a ray r_2, reflected from the conducting plane.

The presence of multipath propagation introduces some problems in the calibration procedures and requires precise information as regards to the parameters of the conducting plane (e.g., earth) within the frequency range at which the calibration is performed. At frequencies above 30 MHz, for horizontal polarization, antennas at the same heights, with elevation angles of $\gamma < 45°$, the reflection factor $\Gamma = -1$ is usually assumed. The same assumption may be accepted when the antennas are placed over a terrain covered by an artificial conducting medium, e.g., metallic mesh. Without regard to the above, before starting calibration procedures, even at a site of previously measured ground parameters, checking the ground conductivity and the reflection factor is suggested before any new series of measurements starts. In order to eliminate any influence of variations in the ground parameters, for instance, due to changing weather conditions, and to assure full repeatability of the measurements, covering the site with a metallic, welded mesh made of non-magnetic material is advised.

The main disadvantages of the STA method in calibration applications are as follows:

- Sensitivity of the method to ground parameters when measurements are performed at an OATS.
- Dependence of the input impedance of the STA on the ground parameters.
- Susceptibility of the OATS to external EMF, especially from local BC and TV stations. This may require checking the calibration frequency range for EMI at a frequency at which calibration is performed. In addition, it may be appropriate to increase the radiated power in order to improve the signal-to-noise ratio sufficiently. However, the latter may cause increased radiation from the set and EMI to other services.
- Errors in measurement of the current exciting the transmitting antenna.
- Susceptibility to the presence of reflecting objects, giving rise to multipath propagation between the antennas.

3.1.2   Calibration with the substitution method

Calibration of E-field meters equipped with dipole antennas using the substitution, or SRA, method is presented in Figure 3.2. An arbitrary transmitting antenna is placed at a distance d from the standard receiving antenna A_r, and the separation, d, should be selected in such a way that the antenna A_r is illuminated by a homogeneous field. Contrary to the STA method, the polarization of the field, to a first approximation, may be an arbitrary one. After measurement of the E-field using SRA (calibration of the field), the SRA is substituted by an antenna under test A_c of a calibrated device. The accuracy of the procedure requires accurate replacement of the antennas. Usually the transmitting antenna and the receiving antenna are placed, as in the case of STA, parallel to one another and with their centers on a common plane. In the case of STA, such placement was required. Here, it is suggested due to EMF homogeneity in the plane of receiving antennas, better polarization matching and stronger signals at the receiving antennas, and, especially, easier placement of the antennas and repeatability of procedures and, as a result, better accuracy of calibration.

The value of the EMF intensity is established here on the grounds of the parameters of the SRA and the electromotive force (emf) measurement at its input or voltage or current measurement when the antenna is loaded.

The main inconveniences of the method may be listed as follows:

- When measuring at an OATS, or in the presence of other conducting objects, the input impedances of the standard receiving antenna and of the calibrated one are affected by those objects. This leads to a difference in measurements as compared to the free-space conditions from antenna to antenna.
- Difficulties with precise replacement of the antennas and keeping the geometry of the propagation constant during any replacement.
- Susceptibility to external EMI is much higher compared to the STA method, especially when the emf at the SRA is measured using a wideband detector.
- Influence of the transmitting antenna excitation power fluctuations and changes of the propagation conditions during measurements.

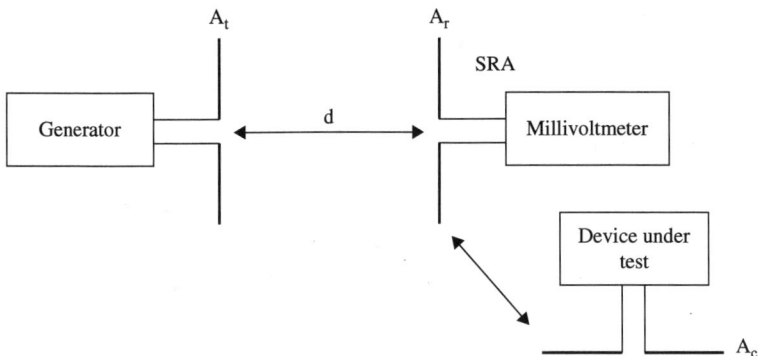

Figure 3.2   Calibration of a meter with a dipole antenna using SRA method

3.1.3   Calibration of meters with whip antennas

Although the calibration of meters with dipole antennas is of primary consideration, we include here a short presentation of the calibration procedure for meters with whip (monopole) antennas. At frequencies above 80–100 MHz, for E-field measurement, resonant antennas are usually used. Below this range, nonresonant dipoles are preferred. At frequencies below 30 MHz, for far-field E-field measurements, meters with whip antennas are generally used. Unlike dipole antennas, which allow the measurement of the EMF of arbitrary polarization, whip antenna use is limited to vertical polarization. Resonant, or even nonresonant, dipoles are usually placed on a tripod and connected to a meter by a coaxial cable during measurements. The whip antenna may be placed directly at the meter or, if tripod and connection cable are in use, a tuning unit is placed at the tripod, and the whip is connected to the tuning unit. The tuning unit or/and input preamplifier is necessary here in order to match the antenna to wave impedance of the feeder, and thus improve the efficiency of the system and, as a result, its sensitivity. The similarity of the two approaches is in the E-field measurement, while the difference is in the symmetry of the dipole antennas and the nonsymmetry of the whip ones.

As in the above procedures, it is also possible to use the STA method as well as the SRA method. The main disadvantages, and the main source of calibration error, are:

- the sensitivity of the calibration procedure to the parameters of the medium above which the procedure is applied,
- placement of the calibrated antenna, and
- presence and location of interconnecting cables.

In order to improve the accuracy the procedure, it is usually performed at places covered artificially by a conducting medium. In the case of the STA method, the geometry of the calibration is shown in Figure 3.3. The standard transmitting antenna A_t, of length l, fed similarly to Figure 3.1, is located at distance d to an antenna A_c of a calibrated device. The two antennas are placed parallel, directly above a perfectly conducting medium.

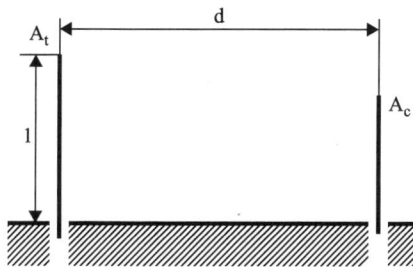

Figure 3.3   Calibration of a meter with a whip antenna

If the length of the STA, the distance between the antennas, and the current exciting the STA are known, the value of E-field at the calibrated antenna A_c may be estimated with (3.1) [12]:

$$E = \frac{15Ik}{\sin^2 kl} \left\{ \frac{2}{3} k^4 l + jk \left[4\sqrt{d^2 + l^2} - 3d - \sqrt{d^2 + 4l^2} \right. \right.$$

$$\left. \left. + 1 \cdot \ln \frac{\left(\sqrt{d^2 + 4l^2} + 2l \right) \left(\sqrt{d^2 + l^2} - 1 \right)}{\left(\sqrt{d^2 + 4l^2} - 2l \right) \left(\sqrt{d^2 + l^2} + 1 \right)} \right] \right\} tg^{-1} \frac{kl}{2} \qquad (3.1)$$

where: k – propagation constant, $k = 2\pi/\lambda$.

Equation (3.1) was introduced with an assumption of a high slenderness ratio in the transmitting antenna and, as a result, the assumption of a sinusoidal current distribution along the antenna. The assumption is generally valid, as usually this is applied to thin whips much shorter than the wavelength. For thin antennas, more complex formulas are to be used.

A similar approach, based upon (2.17) and (2.18) and then averaging the EMF along a calibrated antenna, taking into account mirror reflections in the ground, as shown in Figure 3.4, was applied by the NIST. The inaccuracy of the procedure was estimated at the level of ±1.5 dB within a frequency range of 30 kHz to 300 MHz [27].

The above presentation of the calibration procedures was based upon the STA concept. Now let's consider the same set-up as shown in Figures 3.3 and 3.4 with the substitution method applied. The substitution procedure is exactly the same as that applied when the dipole antennas are being calibrated. The matter is in calculation of the effective height of the standard receiving antenna (h_{eff}) and in emf (e_A) measurement at its input. Then the E-field intensity is calculated using the formula:

$$E = \frac{e_A}{h_{eff}} \qquad (3.2)$$

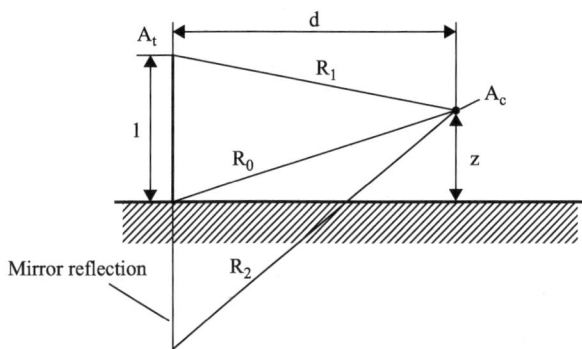

Figure 3.4   Geometry of propagation for a whip antenna calibration

The calibration set-up presented in Figure 3.3 is idealized as both the antennas are placed directly over a perfectly conducting plane. In practice, both antennas are placed on a tripod and are connected to a matching box or tuning unit. Such a configuration substantially limits the accuracy of the procedure and makes it very sensitive to the altitude over the ground at which the antennas are placed, the position of the antennas in relation to other equipment, wiring, ground conductivity, etc. In order to limit the influence of the latter factors, an artificial conducting ground is sometimes applied. The use of the ground during calibration may improve the calibration accuracy, but in real conditions, during field measurements, the use of such a means is rather impossible to imagine. Another factor that is difficult to maintain is the configuration of wiring between calibration and field measurements. Even under the best conditions, the procedure is very inaccurate, and the measurements with the whip antenna are inaccurate as well (accuracy of measurement cannot exceed that of the calibration). The only advantage here is the possibility to directly measure E-fields at low frequencies. For more accurate E-field measurements, meters with loop antennas are used. However, although the meters are calibrated in E-field units, we must remember that a set is sensitive to H-field only (simultaneous sensitivity of the set to the E-field is unwanted and usually results from the so-called "antenna effect," i.e., sensitivity of a loop antenna to the E-field as well). It should be remembered that a presentation of results of E-field measurement in the near field made using a loop antennas is a gross error and may only confirm that the measuring team is inexperienced in EMF metrology!

The equivalence of the E- and H-fields in the far field allows the use of a substitution method in the calibration where the source of EMF may be, for instance, a BC station. Its field is primarily measured with a loop antenna meter, and then it is substituted by the whip antenna set.

The main disadvantages of the whip antennas calibration methods may be summarized as follows:

- Methods are very sensitive to the set configuration and geometry of propagation.
- Very sensitive to the presence of members of the measuring team, wiring, and other conducting objects in the area of calibration.
- Difficulties with matching a short standard antennas (maximal length usually does not exceed 2 m) to its load and especially to the set exciting source and, as a result, relatively low levels of generated field, even compared to the dipole sets presented above; the effect is especially troublesome at the lowest frequencies (kilohertz range).
- Complex and inaccurate formulas describing the field in the case of the STA method.
- Remarkable disagreement of calibration accuracy and, then, the accuracy of measurements.

In order to reduce these disadvantages, the authors propose a new method of calibration with the use of an H-field source. The method is presented in section 3.2.3.

The methods of calibration presented in section 3.1 are designated for low-level EMF measurements in the far field. Such a metrology is typical for EMI measurements, propagation studies, and radiation pattern measurements on an OATS and similar environments. EMF levels generated here are from the microvolt range to a maximum of single volts per meter, i.e., within the range of ± 60 dBm/m. This range fully meets requirements as calibrated meters are of a similar sensitivity range. Their use for stronger field measurements is not advised because of finite attenuation of their screening and possible penetration of the measured field into the meter via other means than their antenna.

3.2   Calibration of meters with loop antennas

Meters equipped with magnetic-type antennas (sensors) as loop antennas and ferrite rods are in wide use for E-field measurements at frequencies below 30 MHz. As mentioned above, the measurement is fully correct in far-field conditions where the mutual relation of the E- and H-fields is constant and may be recalculated one to the other, and the "proportionality coefficient" is just the intrinsic impedance of free space (Z); their mutual relation is expressed by the formula

$$E = Z \times H \tag{3.3}$$

The application usually operates at low-level H-fields, and similar levels are required for antenna and device calibration. Another wide area of H-field sensor application are measurements related to labor safety and environmental protection, where most measurements are performed in the near field. Such near-field measurements involve much higher field levels, and the use of H-field sensors is irreplaceable, as the two field components (E and H) must be measured separately. Apart from quantifying unwanted exposure by sanitary or environmental protection services, this metrology creates opportunities for a wide range of basic studies in medicine, biology, and technology. First, susceptibility studies of *in vivo* and *in vitro* biological objects and a variety of devices and materials require a much higher frequency range, from magnetostatic fields up to and beyond 1 GHz, and with remarkably higher H-field intensities. The "low-level" metrology is based, in the majority of cases, on a selective measurement, limited to single frequency, or a maximum of one communication channel, This is true even if spectrum analyzers are used to visualize any frequency fringe within a measured spectrum of frequencies. In the case of "high-level" fields, wideband devices are usually used. Contrary to the selective meters whose frequency response ensures separation to any unwanted frequency fringes, here the shape of the frequency response out of the measuring frequency band may be non-controlled, and the sensitivity of a device out of the band may exceed that within the band. This leads to gross errors during measurements in a dense electromagnetic environment with uncontrolled emissions. The effect may be especially pronounced when an H-field sensor is constructed from a small loop antenna, and the field appears at frequencies where the size of the antenna is comparable with the wavelength and the antenna becomes a "resonant one." In order to ensure the required quality of measuring equipment during its manufacture and during its periodic control, the frequency response should be checked at much higher frequencies than the meter's upper corner frequency. The protection standards usually require H-field measurements up to 300 MHz or above; however, the effects described require the possibility to generate much higher H-fields. The world's first H-field standard working at frequencies above 30 MHz was designed and completed in cooperation with the NIST by the authors [41].

3.2.1 Calibration with the standard field method

The configuration of the set-up for the standard field method is identical to the case of the dipole antenna calibration. A standard transmitting antenna A_t is parallel and coaxially placed at distance d from an antenna under test A_c, in such a way that the vector units of their planes are on the same line, as shown in Figure 3.5. Average H-field intensity in the plane of the calibrated antenna is calculated on the basis of the antennas' sizes, the geometry of propagation, and the measurement of the transmitting antenna exciting current, as was discussed in section 2.5.

Figure 3.6 shows a laboratory set-up for calibration with the standard field method applied by the authors. The figure shows a bench at which is placed a

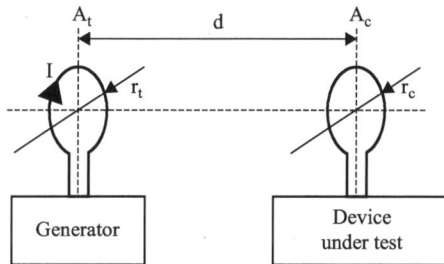

Figure 3.5   A calibration with the standard field method

Figure 3.6   A laboratory set-up for H-field antenna calibration with the standard field method

calibrated loop antenna with its tuning unit and a standard transmitting antenna behind it. At the top of the STA, a thermocouple for the antenna's current measurement is connected. A power source may be seen on the table, on the right side.

The transmitting antenna and the calibrated one are placed at a distance not less than 3d away of any other objects; this makes it possible to assume that the field coupling between the antennas is undisturbed. Necessary leads are behind the antennas, and it can be shown that, set out this way, their role in the field disturbances and the accuracy reduction may be neglected.

The main practical disadvantage of the method is the problem of matching a small loop antenna to a standard output impedance of a power source exciting it. The problem is negligible when low-level fields are generated. However, it is of primary importance when high field strengths are to be generated and maximal efficiency of radiation is necessary. This phenomenon limits maximal intensities to the level of several tens of A/m. When higher intensities are required, more complex solutions should be considered. Some problems may appear with an influence of external EMF during far-field meter calibration under standard laboratory conditions. Although the effect is not very troublesome (as selective meters are calibrated), the effect may be absolutely eliminated when calibrations are performed in a screened chamber.

The main advantage of the set-up is its stability. In practice, the set-up is insensitive to the presence of any objects at comparatively small distances from it. This is a very important advantage, especially when compared to the whip antenna calibration, as presented in section 3.1.3.

As has already been mentioned, in order to make it possible to check the frequency response of a meter under test over a wide frequency range, it is necessary that the range of the H-field standard be much wider than that of the calibrated antenna. Moreover, the field should be as strong as necessary for the calibration. In order to assure such a possibility, with the use of comparatively simple and inexpensive solutions, the authors proposed, using the STA method, a tuning of the standard transmitting antenna to resonance, as shown in Figure 3.7 [22].

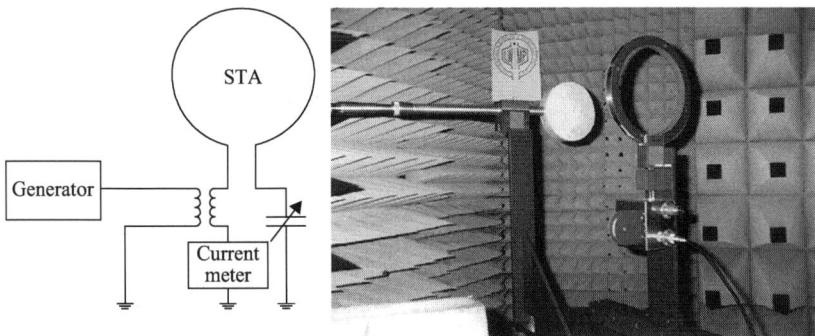

Figure 3.7   Schematic diagram of an STA tuned to resonance (left) and authors' set-up during calibration (right)

The presented set-up contains several multi-turn loops, for several frequency subranges, and works within a frequency range of 30 kHz to 30 MHz. It allows H-field intensities up to 10 A/m at distances up to 50 cm when excited from a 5 W power source. The concept is based upon the known rule that, in series resonance, the current flowing in the circuit increases Q times (where: Q – quality factor of the circuit). The set is almost universal in use. However, it requires caution during calibration, because of relatively strong E-field component accompanying the H-field, which may completely disturb the results of calibration if the calibrated meter (probe) is sensitive to the E-field as well. The simplest way of checking the sensitivity is rotating a calibrated device through 180°; during such a procedure the indications of the device should remind unchanged. The authors' set-up shown in Figure 3.7 was placed in an anechoic chamber, although it is not "anechoic" within this frequency range. The reason was the limitation of the radiated EMI as the chamber is well screened.

3.2.2 Calibration with the substitution method

The steps and procedures for calibrating magnetic-type antennas using the standard receiving antenna method are identical to those described in section 3.1.2 in relation to dipole antenna calibration. The set-up for antenna calibration by the SRA method is shown in Figure 3.8.

At an arbitrary point in space, a standard receiving antenna A_r is placed where an acceptably uniform magnetic field exists. It is possible to generate maximal H-field intensities and allow simultaneous calibration with the use of the STA and SRA methods. For the placement of a transmitting antenna A_t, the SRA is suggested, identical to the previously presented case using the STA method (Figure 3.5). A configuration applied in routine calibrations by the authors is shown in Figure 3.9. The figure shows a universal bench for H-field calibrations, at which is placed a transmitting, multi-turn loop (right) and an H-field probe (calibrated device) after replacement with an SRA (left).

The electromotive force (emf), e_A, induced by an H-field in a plane loop antenna of surface S_r in a free, nonmagnetic medium is given by Faraday's law:

$$e_A = -\frac{d}{dt}\int_{S_r} \mu_0 H \, dS_r \qquad (3.4)$$

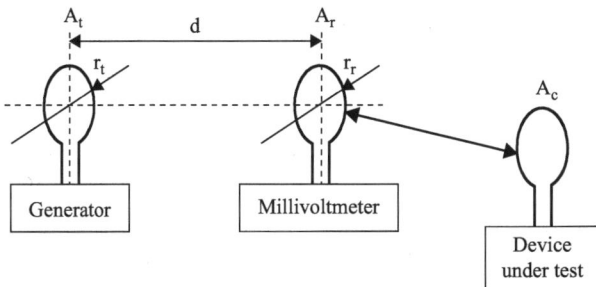

Figure 3.8   Calibration H-field antennas by the SRA method

Figure 3.9   A laboratory calibration set-up for H-field antennas calibration by the SRA method

where:

μ_0 – permeability of free space.

If it is assumed that the antenna is illuminated by a monochromatic wave, the field distribution at the surface of the antenna is uniform, and the vector H and the unit vector of SRA are parallel, on the basis of (3.4) we may calculate the H-field (H_{AV}) averaged by the antenna as:

$$H_{AV} = \frac{e_A \, 10^7}{2 f n S_r} \qquad (3.5)$$

where:

n – number of turns and f is the frequency in Hz.

Equation (3.5) allows the use of any loop antenna in the role of the standard receiving antenna. However, circular loops are most commonly used. Averaged H-field strength at the plane of the SRA, for the selected type of antenna, whose parameters (diameter, number of turns) are well known, is most often calculated using (3.5). Usually the calculations are checked experimentally. After the H-field calibration, the SRA is replaced by an antenna under test with the assumption that the field remains unchanged, without regard to any external conditions that could be changed in the meantime and assuming that existing couplings between the transmitting antenna and both the receiving ones are identical. During the procedure, one must remember the difference between the field averaged by the SRA and the antenna under test when their diameters are different. This case was discussed in section 2.5. The substitution method is used mainly for low-sensitivity meter calibration; thus, quite high H-field intensities are necessary.

The most important disadvantage of the method is related to the latter requirement; similar to the STA method, the generation of strong fields creates technical challenges. In order to even partially solve the problem, potentially large multi-turn loops are used. However, such a solution limits the frequency range in which the antenna can work to a maximum of a single decade. As a result, within a frequency range from kilohertz frequencies to 30 MHz, several different loops have to be used.

The standard field method may be applied both for sensitive far-field meters and much less sensitive near-field ones. The maximum obtainable H-field intensities, reaching several hundreds of amperes per meter, are limited by available power sources, by possibilities to measure a current exciting the antenna, and by safety considerations in order not to damage or even burn the antenna. Up till now, the SRA method has been mainly applied to near-field probes and meter calibration, or as a method that made it possible to prove the STA method. Instead of quite large, multi-turn SRAs, the authors propose an application of relatively small, active loops that make it possible to use the method for far-field meter calibration as well. In this case, the SRA is loaded by a wideband amplifier of relatively high input impedance, and its output voltage is measured by a selective milivoltmeter.

3.2.3   Calibration of meters with whip antennas

Section 3.1.3 presented a traditional approach to whip antenna calibration, listing factors that reduced the accuracy of the method. The authors took into account that the magnetic field source (a loop antenna) has two spatial components of H-field [given by (2.28) and (2.29)] and single one of E-field (E_ϕ). This suggests the possibility of using a loop antenna as a source for a standard electric field as well. The idea is presented in Figure 3.10, which shows a coplanar standard loop antenna of radius r_t and, at distance d from it, a calibrated whip antenna of length l [42].

The E-field component of a small loop antenna E_ϕ, which appears simultaneously with two H-field components given by (2.28) and (2.29), is:

$$E_\phi = \frac{I \cdot S_t}{4\pi \cdot R_0^3} Z \sqrt{1 + k^2 R_0^2} \sin \Theta \qquad (3.6)$$

If account is taken of the E-field averaging at the whip, as given by (2.22), and assumptions similar to those accepted when (2.37) was introduced, we will have,

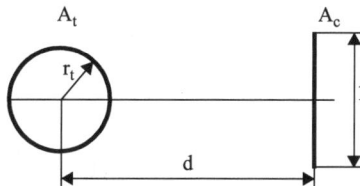

Figure 3.10   A whip antenna calibration with a standard transmitting loop antenna

for $1 < d/4$ and $r_t < d/4$, with an accuracy to 1%, the averaged E-field (E_{AV}) at the length 1 of the calibrated whip, expressed in the form:

$$E_{AV} = \frac{\pi r_t^2 Z I}{2\lambda d D} \sqrt{1 + k^2 D^2} \tag{3.7}$$

where:

$$D = \sqrt{d^2 + r_t^2 + 1^2} \tag{3.8}$$

The structures of (3.8) and of (2.37) are identical. Although the radiation efficiency of the loop is proportional to frequency, the efficiency is much better compared with a transmitting whip, especially at the lowest frequencies. In order to illustrate the frequency dependence of the E-field, Figure 3.11 shows E-field run versus frequency at distance $d = 1$ m from a circular loop antenna of 20 cm diameter when excited by sinusoidal current of 1 A intensity.

In order to prove agreement of the E-field strength generated using the proposed method compared with another one, a test was done with the use of a transfer standard in the form of short, symmetric dipole antenna. Comparative calibration was performed using a TEM cell. Results of the comparison are set out in Table 3.1.

The results presented in Table 3.1 show a disagreement of calibration results between the use of the loop antenna E-field standard and the TEM cell. However, the difference does not exceed ±5%, which may be assumed to be fully acceptable.

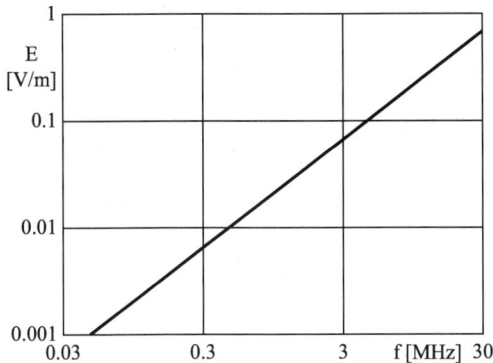

Figure 3.11   E-field versus frequency of a loop antenna

Table 3.1   Results of comparative measurements (in V/m)

f (MHz)	30	21	14	7	3.5	1.7
E (TEM)	2.3	1.80	1.24	0.58	0.281	0.135
E (loop)	2.5	1.72	1.16	0.56	0.269	0.126

Evident advantages of the proposed method may be summarized as follows:

- Identical set-up for E- and H-field generation.
- Simple formulas.
- More easily measurable excitation.
- Much higher E-field intensities are obtainable when compared with whip transmitting antennas.
- Relatively insensitive to the presence of any other objects in its neighborhood.
- Better accuracy.

3.3   Calibration of meters with directional antennas

Nondirectional antennas, with the exception of active probes of spherical directional pattern for near-field EMF measurements, do not exist. By "directional antennas," we mean antennas of directivity exceeding that of dipole antennas, for both electric and magnetic fields. EMF meters with directional antennas are used, usually, at frequencies above 300 MHz: where sizes of the antennas, comparable to or exceeding the wavelength, are acceptable, especially for outdoor measurements. The most common types of directional antennas are horn antennas, log-periodic antennas, Yagi-Uda antennas, multi-dipole (active and passive) arrays, spiral ones, etc. (Figure 3.12).

Especially at frequencies below 1000 MHz, it is possibile to apply directional antenna calibration in the same way as for dipole antennas (see section 2.1). Above that frequency, construction of dipole antenna standards is troublesome because of short wavelengths and resultant technical difficulties with designing thin dipole antennas. Moreover, because the gain of dipole antennas is low, this may cause problems with high-gain antenna calibration. Thus, below 1 GHz, the role of standard antennas usually applies to horn antennas.

At frequencies above 300 (1000) MHz, other measurements are usually used, i.e., power density (power density flux) S, and contrary to meters applied at lower frequencies, where antennas (probes) should be usually calibrated together with a (voltage) measuring device, here power at an input of a (standard) receiving antenna is measured, and because of full impedance matching of the antenna and a measuring device, the calibration may be performed using an arbitrary (power) meter.

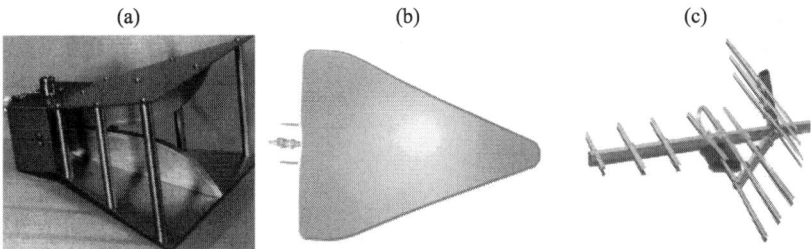

(a)   (b)   (c)

Figure 3.12   Directional antennas: (a) wideband horn, (b) log-periodic antenna, (c) Yagi-Uda antenna

Directional antennas are usually of high gain. Thus, during their calibration, it is possible to neglect the influence of multipath propagation, the presence of conducting objects and their possible effect on the antennas' input impedance, external interference, and other confounding variables. Of course, it does not change the fact that an anechoic chamber is recommended, particularly for its climatic advantages in relation to an OATS.

Similarly, as with dipoles and loop antennas, it is possible to apply both the STA and the SRA methods. Both methods are based upon the use of a standard horn antenna that can be used as a receiving as well as a transmitting antenna, and the most important parameter of the antenna is its gain G. In the case of the STA, the power density S at distance d from a standard transmitting antenna is given by:

$$S = \frac{P_t G_t}{4\pi d^2} \tag{3.9}$$

where:

G_t – power gain of the STA,
P_t – power fed to the antenna.

The formula is valid for plane wave conditions; i.e., the following condition should be fulfilled: $d \gg A^2/\lambda$ (where: A – aperture surface of the largest antenna applied in the procedure). If the condition is not fulfilled, it is necessary to take into account appropriate correction factors, which may be estimated theoretically or experimentally.

For the SRA method the power density at the surface of the receiving antenna is given by:

$$S = \frac{P_r}{A_r} = \frac{P_r}{G_r} \frac{4\pi}{\lambda^2} \tag{3.10}$$

where:

P_r – power measured at the input of the SRA,
A_r – effective surface of the SRA,
G_r – power gain of the SRA.

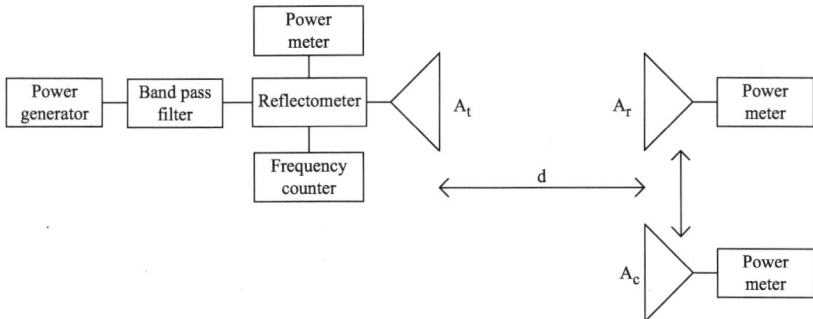

Figure 3.13   A block diagram of the directional antennas calibration system

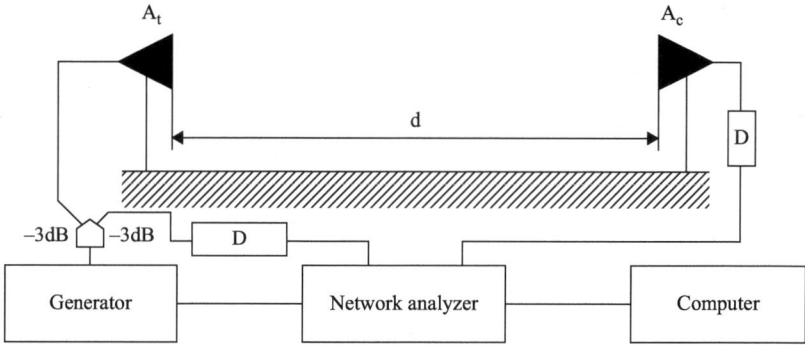

Figure 3.14   Block diagram of the calibration system applied by the authors

A block diagram of the calibration system is shown in Figure 3.13. The system contains, at the transmitting side, an STA (A_t) fed from a power source through a bandpass filter and reflectometer. The reflectometer allows an excitation measurement of the antenna, and the wavemeter may be necessary if the power source does not ensure accurate frequency readings, as is often the case in power generators. A bandpass filter is especially necessary when a power source can produce nonlinear distortions of the signal fed to the antenna (the role of harmonics is discussed in section 5.3.3). Parallel to the STA and coaxially to it, at distance d, is placed an SRA (A_r) loaded with a power meter, which is then substituted by a calibrated antenna A_c loaded by a power meter, a device under test, or any similar meter. The system is of universal character and allows for both the STA and the SRA method.

A block diagram of an automated calibration system used by the authors is shown in Figure 3.14. The system is based upon the STA concept and contains an STA (A_t), fed from a generator through a power divider that allows the STA excitation measurement. Both the exciting power and the power from the antenna under test (A_c) are fed through detectors to a network analyzer.

The main disadvantage of this set-up is the need to precisely orient the antennas. This was not as important in the case of dipole or loop antennas because of their relatively low directivity. A further inconvenience is the need to correct for the standard antenna parameters, due to their variations with frequency, when the set-up is retuned from one frequency to another. However, the latter may be eliminated as the procedure may be fully automated, for instance, as shown in Figure 3.14.

It is worth remembering here that the procedures presented are designed, mainly, for calibration of antennas applied in EMF metrology. Measurements of large antennas and antenna systems, in particular radiation pattern measurements in the near field, applied in radio communications, radiolocation, etc. require a special approach [38], and these cases are not taken into considerations here.

3.4   Meter calibration with the use of guided waves

An application of guided wave systems for calibration procedures is very convenient, as they ensure almost full insulation of an object under test against the external electromagnetic environment and vice versa. Moreover, there exists a possibility to generate relatively strong EMF with the use of comparatively low exciting power. The main disadvantage of the systems is the limited space within them. It allows investigation of objects whose sizes are several times smaller compared to the sizes of systems applied. This limitation is more and more important as the frequency increases. Moreover, much stronger mutual coupling between the system and an OUT is noted as compared to previously presented procedures. This coupling may affect the results of an investigation, or even produce false results.

3.4.1   Calibration in the field of a plate capacitor

Construction of a calibration system using a parallel plate capacitor is similar to that for the guided wave approach, and the advantages and disadvantages of the capacitor are similar to those of the guided wave system. However, there are no guided waves in the capacitor and radiation of EM energy into its surroundings is substantial. It may be compared to an open-ended strip line.

A block diagram of a calibration system with a plate capacitor is shown in Figure 3.15. A voltage from a generator is fed through a symmetrizing transformer (ST) to the plates of the capacitor. At the input of the capacitor, its exciting voltage is measured by a symmetric voltmeter. The E-field intensity within the capacitor is given by:

$$E = \frac{V}{D} \qquad\qquad (3.11)$$

where:

D – distance between the capacitor plates,
V – voltage between the plates.

Equation (3.11) is fully correct if the E-field distribution inside the capacitor is uniform. This assumption is correct when the plates are comparatively large in comparison to the distance between them, and the field is not disturbed by a calibrated antenna or any other OUT. Moreover, the maximal sizes of the plates must be much smaller compared to the shortest wavelengths at which it is to work, due to the standing waves at the plates. It is usually assumed that the maximal sizes of the capacitor should not exceed 0.1λ. In order to minimize couplings between the

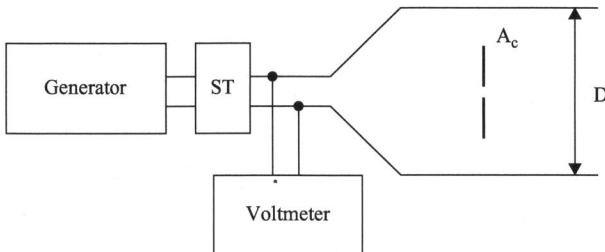

Figure 3.15   A EMF standard with a plate capacitor

Figure 3.16 A view of the plate capacitor EMF standard

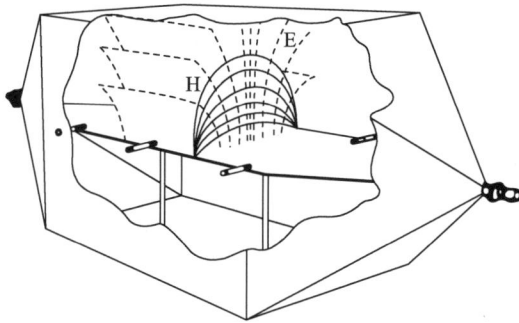

Figure 3.17 E- and H-field distribution in a TEM cell

capacitor and its surroundings, as well as the couplings with an OUT and its leads, it is recommended that the voltage fed to the capacitor be symmetrical. If the system is nonsymmetrical, different readings of a calibrated meter (probe) may be observed when it is rotated through 180° during calibration.

A set-up with a plate capacitor used by the authors is shown in Figure 3.16. A calibrated EMF meter placed on a dielectric sheet may be seen inside the capacitor.

3.4.2 EMF standards with a traveling wave line

In any standards based upon traveling waves, it is assumed that the EMF distribution is identical to static conditions. This assumption is correct if the standard works with a basic mode, the presence of higher modes may be neglected, and the field inside the device is not distorted by the presence of an OUT.

One of the most popular solutions to the traveling wave EMF standard is a part of a symmetric strip line. Because of the quasi-TEM field structure within the line it is called a TEM cell; previously it was called a Crawford cell, after the original author [11]. A view into a TEM cell and E- and H-field distribution inside it is shown in Figure 3.17.

The internal plate of the line, supported by dielectric supports, is placed between four walls. This makes a fully screened TEM cell; in open cells there are

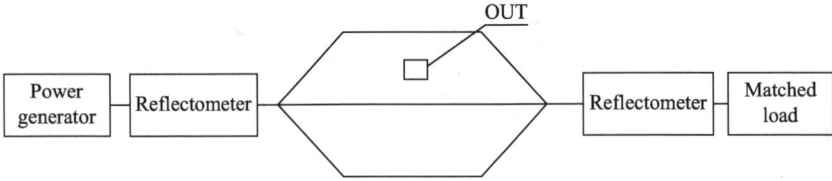

Figure 3.18   A TEM cell as an EMF standard

no sidewalls. The role of the sidewalls is only to eliminate any coupling with the surrounding environment. The E-field between the center conductor and the side-walls is reminiscent of the parallel-plate capacitor, and it leads to a similar relationship between the excitation of the cell and the E-field within it:

$$E = \frac{V}{D} = \frac{\sqrt{P_e Z_c}}{D} \qquad (3.12)$$

where:

P_e – power exciting the cell,
Z_c – wave impedance of the cell,
D – distance between the center conductor and a sidewall.

As may be seen from Figure 3.17, both ends of the cell are matched to the coaxial connectors and coaxial wiring by matching transformers. Apart from mechanical matching, they match the wave impedance of the cell to the standard impedance of the whole system (usually 50 Ω), i.e., wave impedance of the feeding devices from one side and loading the cell from the other. In order to have good matching, as well as negligible influence of the transformers upon the field homogeneity inside the cell, the transformers should be long enough; however, their length limits the bandwidth of the system. A block diagram of a standard set-up with a TEM cell is shown in Figure 3.18. The system contains a generator connected to the cell through a reflectometer that enables a measurement of both traveled and reflected waves (necessary for the power balance measurement when power absorbed in the cell is of concern) and matched load connected to the cell by a similar set-up for measurements of both traveled and reflected waves.

Although the E-field in a TEM cell is calculated similarly to that in a capacitor, it is impossible to precisely calculate the H-field in the capacitor system. Due to the presence a quasi-plane wave in the cell, it is possible to use it as an H-field standard as well. The relation between the field components within the cell is identical to that for the free space given by (3.3); will present it in a slightly modified form:

$$H = \frac{E}{Z} = \frac{\sqrt{P_e Z_c}}{ZD} \qquad (3.13)$$

where:

Z – intrinsic impedance of free space.

In summary, the most important advantages of the cell include:

- the ability to generate quite strong fields,
- isolation from the external environment,
- simple relationships describing the field within the cell,
- possibility to automate procedures,
- application over a wide range of frequencies; in practice from a static field to an upper corner frequency when a higher mode starts to appear.

The disadvantages include strong couplings between the system and an OUT, mismatching and distortions of EMF distribution within the cell caused by the OUT, and limited usable volume in the cell. The significance of these negative phenomena increases with the frequency increase.

3.4.3 A waveguide as a standard EMF source

The length of a rectangular waveguide may be used for electrically small EMF probe calibration, particularly when operating in the basic mode. Waveguides such as these can also be used for other studies, such as biomedical research on micro-organisms. As in the case of a TEM cell, there is a possibility of relatively precise measurement of the waveguide excitation as well as a calculation and/or measurement of the field distribution in the waveguide. Thus, precise field strengths may be calculated within the waveguide, again, particularly in the basic mode. Because of the wavelengths at which the approach may be applied, the area in which the field may be assumed homogeneous is relatively small, and this limits the sizes of any OUT.

In order to illustrate the above statement, Figure 3.19a shows a cross section of a rectangular waveguide working within the 10 GHz band with basic mode TE_{10}, used for the calibration of EMF microprobes. The TE_{10} mode was selected because of minimization of the field disturbances inside the waveguide due to a slot cut-off as well as the calibrated probe insertion in the waveguide [5].

In the case presented, a microprobe of 0.6 mm was calibrated in the wave-guide. The sizes of the probe and its casing required a slot 3 mm wide to freely

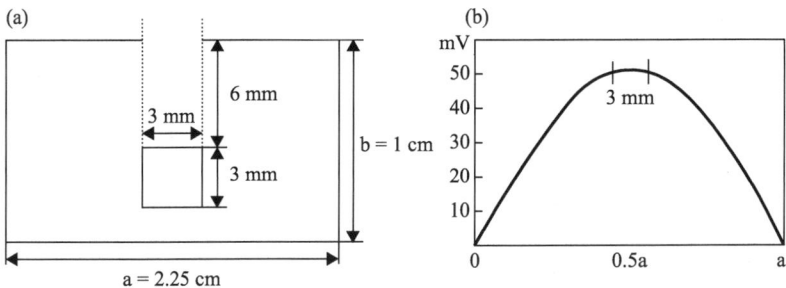

Figure 3.19   A cross section of a waveguide: (a) indicated sizes of the waveguide and an area 3 × 3 mm where EMF may be assumed to be homogeneous; (b) E-field distribution in the waveguide

immerse the probe in the waveguide. Figure 3.19b shows an area of 3 mm in the wave maximum, where the homogeneity of the field is no worse than ± 1 dB. The E-field strength E in the maximum for the TE_{10} mode is:

$$E = \sqrt{\frac{P_e}{ab} \frac{\lambda_g}{\lambda} Z} \tag{3.14}$$

where:

P$_e$ – power fed to the waveguide,
a, b – sizes of the waveguide (as in Figure 3.19a),
λ_g – wavelength in the waveguide:

$$\lambda_g = \sqrt{\frac{\lambda}{1 - (\lambda/2a)^2}} \tag{3.15}$$

Agreement between the comparative calibrations performed in a waveguide with the use of system as shown in Figure 3.19, and in an anechoic chamber was estimated at $\pm 18\%$.

The use of the method was stimulated by the necessity to have a tool that would work at microwave frequencies, and this ability is the main advantage of the approach. Another advantage is the possibility of generating strong fields compared with the methods presented in section 3.3, not to mention the good isolation against the external EM environment.

Disadvantages of the method are as follows:

• very small applicable area,
• relatively narrow frequency band in which the system may be applied,
• frequency dependence of field intensity and distribution in the waveguide,
• alternating relation of E/H in function of frequency and a spot considered.

3.5   Secondary standards and exposure systems

The main purpose of the methods presented above and of systems of standard EMF generation is to obtain precise values for E- and H-field intensity, based on a knowledge of the applied system structure (based on theoretical or experimental approaches), measurement of the system excitation (from a power source in the case of STA or by EMF in the case of the SRA method), and the geometry of propagation. One of the most important considerations was the accuracy of the procedures. Although these methods are accurate, they are usually complex and troublesome in use, which often makes it impossible to automate them. Moreover, in methods using an OATS, it is impossible to generate sufficiently strong field strengths. Another disadvantage is a sensitivity to the external environment and mutual interaction of a standard and a device under test. Thus, the majority of methods can be applied when the dominating requirement is accuracy of calibration. The requirement is not always rigorous, which allows use of simplified methods that are easier in practical procedures. We will call them secondary standards or exposure systems. In the majority of solutions, EMF intensity is measured using a probe (antenna) that was previously calibrated using a primary standard. There are a variety of implementations of these devices, and the following sections will not consider them in great detail as there are almost as many different versions as metrological needs, and the literature on the subject is very extensive. We will discuss factors limiting the accuracy of the standards, and the discussion will be valid for the secondary standards as well as the physical phenomena; factors limiting accuracy are similar almost everywhere.

3.5.1   Chamber methods

One of the most popular methods of EMF generation, especially for EMC experiments, is a concept based upon mode reverberation in a screened chamber. The idea of the chamber, incorporated to the MIL-STD-1377 [after 14], is shown in Figure 3.20.

A transmitting antenna is placed in a screened chamber. The antenna may be loaded with a matched load in order to have a traveling wave in the cabling instead of a standing wave. The antenna is fed from a power source thought to be a matching unit. Because of frequency dependence of the antennas' input impedance, the matching device should be automatically tuned, especially when measurements are automated. Inside the chamber, one or usually more probes are placed for sensing (calibrating) the field intensity within the chamber. Without regard to the wave type in the transmitting antenna at a frequency, the EMF distribution within the chamber is static, and the maxima and minima of standing waves and their presence and location are functions of the relation between the chamber size and the wavelength. In order to have quasi-uniform spatial EMF field distribution in the chamber (not in time!), the chamber is equipped with one or more mode reverberating devices. An OUT is placed in a place where the field homogeneity is estimated as maximal [30].

The sizes of the chambers depend of dimensions of the object that may be investigated in the chamber. The largest chambers are destined for ships, planes, and satellite testing. Frequencies at which the chambers work are from megahertz

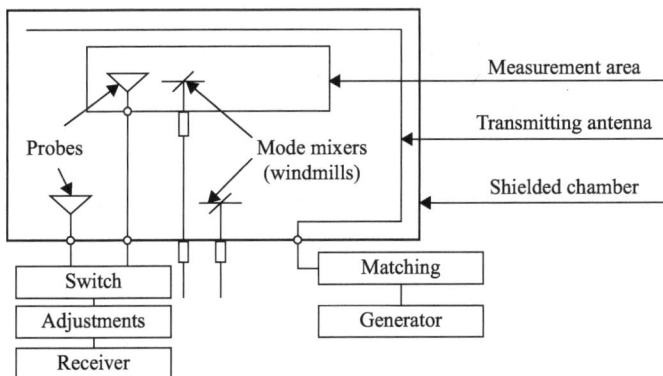

Figure 3.20   A block diagram of a mode reverberation chamber

to high gigahertz ranges, with the minimum operational frequency being inversely proportional to the size of the chamber. The problem of standing waves is overcome by sweeping modes during measurements. In older solutions, field measurements within the chambers used linearly polarized probes. Unfortunately, spatial orientation of the field vectors is unknown. The "stirred" modes may produce a quasi-spheroidal polarization (it should be noted that spheroidal or ellipsoidal polarization does not exist; quasi-spheroidal means that an ellipse of polarization rotates in the space, which makes the field appear to have three spatial components). Thus, omnidirectional probes are preferred in newer solutions, which do not assure the exact field distribution, intensity, and polarization and time variation at a location within the chamber and around an OUT. Moreover, the mutual relation of E- and H-fields varies with time (due to steering) and frequency. In more accurate measurements, E- and H-field probes are applied simultaneously. An example of a chamber at the Technical University of Wrocław is shown in Figure 3.21.

Repeatability of the measurements in a chamber is estimated at ± 6 dB, which can be assumed to be acceptable when taking into account the accuracy of the probe calibration (say ± 1 dB) and the purpose of the measurements, taking into account also the possibility of automating the procedures for quick testing. The measurements are usually done for EMC testing purposes in two large groups:

- susceptibility testing of a variety of electric and electronic devices due to requirements of EMC standards,
- technical testing of devices and systems in order to assure them required protection against external EMI and to limit EMI radiated by them.

The former is usually done in accordance with methods required by EMC standards, and accuracy of measurement is not taken into account; in the latter, the aim is to find reasons and place responsible for leakage of the EM energy into the OUT or outside of it. In the latter, in the case of testing of a single device (e.g., satellites), the aim is optimization of systems and devices for EMC.

Figure 3.21   A view into a mode reverberation chamber (left) and its paddles (right)

Figure 3.22   Shielded room (left) and anechoic chamber (right) at the Technical University of Wroclaw

An example of the chambers constructed and completed at the Technical University of Wroclaw is shown in Figure 3.22.

3.5.2   Examples of TEM chambers

Limited usable volume and the presence of higher modes in mode-stirred reverberation chambers, as well as EMF time variations within them due to mode stirring, simulated further work on designing a more generally applicable test environment. Early research in this area was initiated at NIST in Boulder, Colorado, in the early 1970s. The idea was based upon a combination of a transmission line with a horn radiator. The line, placed above a conducting plate, is loaded at its end by a matched load, and this part of the chamber works within lower frequencies in a way similar to that of the previously presented TEM cells. In this frequency range, the horn plays the role of matching transformer. At the same time, the horn plays the role of a radiator at higher frequencies and, in this frequency range, radiated energy is absorbed by absorbing material that covers the end wall of the chamber. As a result, an EMF structure similar to a TEM wave exists. Then the idea was further developed by Podgorski, Hansen, and others. In order to illustrate the variety of designs, their parameters and possibilities, several of them are presented below.

One of the most active researchers in the field of TEM-like chambers is A. Podgorski. He proposed several designs for the chambers. One of the simplest of his designs, an open EM simulator, is shown in Figure 3.23. The solution includes two plates. The upper plate in this design plays the role of an active line while the other acts as a ground plane. The line is matched at its ends by matching transformers to an excitation source (left) and to a matched load (right). The load here is a single resistor connected between the end of the transformer and the ground plane. The horn is fed by a coaxial line. Then the center conductor of the line is connected to the line via a transformer and the horn simultaneously works as part of the matching system and as a radiator. This version has no absorber at its end. An OUT is placed in the center of the simulator at some height above the ground plane [36].

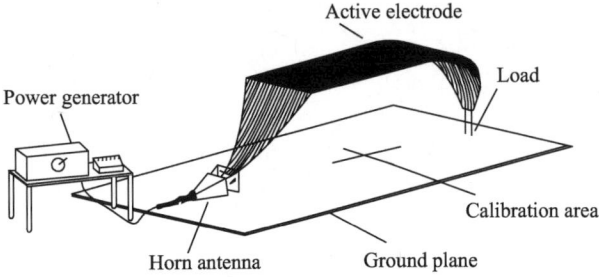

Figure 3.23   Podgorski's open TEM simulator

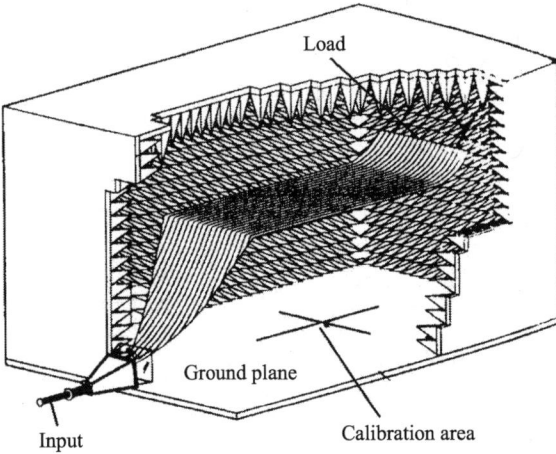

Figure 3.24   Podgorski's closed TEM simulator

Podgorski's open simulator is a very simple and inexpensive design. The main disadvantage of this class of chambers is almost unlimited coupling with the local EM environment, which may be accepted in some simpler applications but not where the external EMF may affect the system. The second proposed design (Figure 3.24) is free of this disadvantage. However, it is much more expensive and may create problems with volume accessible for larger OUTs. Podgorski's closed simulator looks like the open one with two exceptions: A whole system is placed within an anechoic chamber, and the load is dispersed at the width of the center conductor at its end instead of at one point. A wall of the chamber plays the role of the absorbers previously mentioned. The chamber ensures isolation against an external environment, while the dispersed load and presence of absorbers ensure better matching and operation over a wider frequency range [37].

Figure 3.25a presents variations of the E-field intensity versus frequency in Podgorski's simulator, and the H-field is presented in Figure 3.25b. It may be seen from the figures that the simulator makes it possible to have a constant field

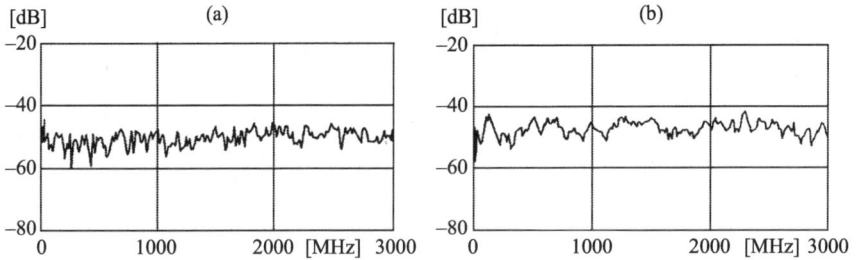

Figure 3.25   Frequency response of the Podgorski simulator: (a) E-field,
(b) H-field

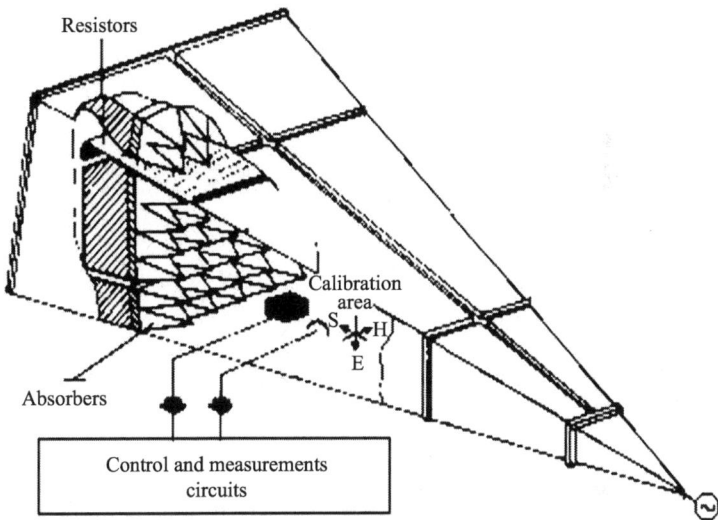

Figure 3.26   Concept of the GTEM cell

intensity with an accuracy to ±6–±7 dB up to 3 GHz. Maximal E-field intensities in the simulator may reach 100 kV/m during a pulse excitation.

A similar concept was developed by Hansen. He proposed several solutions for GTEM chambers. The first solution is similar to a TEM cell limited only to a matching transformer, doubly loaded as in earlier NIST proposals. The absorber, placed at the end part of the cell, plays the role of a nonreflecting load at higher frequencies. Resistive loading, dispersed at the chamber closing plate, behind the absorbers, creates the chamber load at lower frequencies. Every Hansen's chamber is fully screened to ensure good isolation between the space inside the chamber and the external environment. The concept of Hansen's GTEM cell is shown in Figure 3.26. Currently, this solution is one of the most popular tools in EMC-oriented investigations. Different versions of the GTEM cell have been presented in many publications [29], and the cell has been adopted by several EMC standards.

Figure 3.27   GTEM-like cell designed at the Technical University of Wroclaw

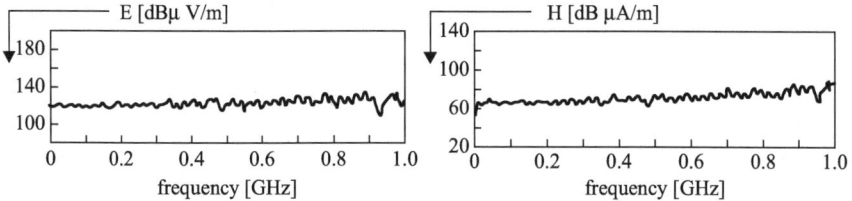

Figure 3.28   E- and H-field levels in the cell shown in Figure 3.27

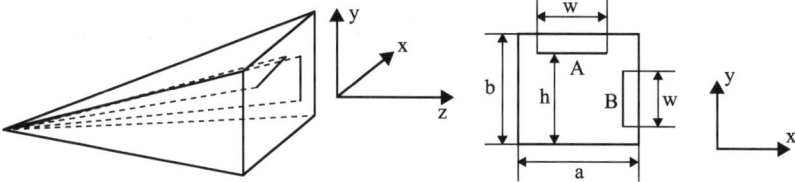

Figure 3.29   A concept and a cross section of a TTEM cell

A GTEM-like cell designed at the Technical University of Wroclaw is presented in Figure 3.27. Measured results of E- and H-field, up to 1 GHz in the calibration area of the cell, when excited with a power of 100 W, are shown in Figure 3.28.

Further development of GTEM chambers led to a TTEM cell (triple TEM cell). Hansen proposed, in one casing, two separate "GTEM" cells. The concept is shown in Figure 3.29. The solution enables simultaneous work at different frequencies or spatial rotating of the field polarization in the cell. If electrodes A and B are excited in quadrature it is possible to have a circularly or elliptically polarized EMF in the cell [24]. The latter may present some difficulties when automated

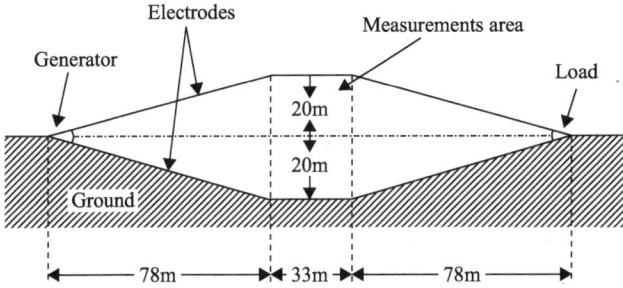

Figure 3.30 A reentrant cell

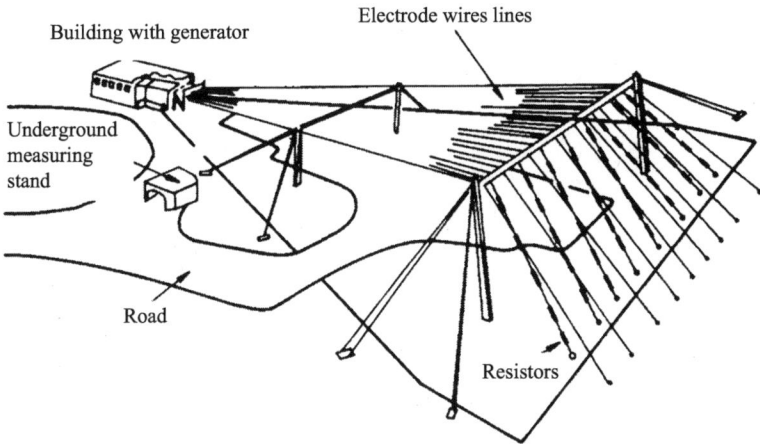

Figure 3.31 A cell with wire electrode

measurements are performed over a wide frequency range, but apart from that it creates a new possibility.

Investigation of large objects may be required, particularly in EMC studies. Several concepts and proposals have been presented in the literature. One of them is based upon natural topographical features. In the case of a large TEM cell, a reentrant was applied to immerse cell walls inside it, as shown in Figure 3.30 [48].

A common feature of the majority of TEM-like cells is their simple construction and relative low cost, even if they are quite large in design. In order to further simplify the design and lower its costs while needing to investigate increasingly large objects, the firm of Brown-Boveri in Zurich proposed a solution in which, instead of plates in the cell walls, wires were used, as shown in Figure 3.31. The ground is covered by a mesh, and every wire is separately loaded by a matching resistor. Such a construction is wind-resistive and easy to build up and break down on an OATS. Such devices, called WTEM cells (wire TEM cells), are known in several versions. One of them was proposed by Carbonini [9].

Without regard to the advantages of the types of the large cells, chambers, and stimulators presented, it is necessary to remember that every one of them works well in a frequency range where the size of the device is less than a wavelength. At higher frequencies, apart from basic mode TEM, higher modes may appear as well. Different solutions are sensitive to the phenomenon to different degrees. However, their upper corner frequency is defined by their size. An approach was proposed to widen the frequency range of the devices beyond the limits and to work with the presence of higher modes as well. An example of such a solution was based upon the use of a standard TEM cell in a way similar to that for mode-stirred reverberation chambers [30]. Paddles were placed in a TEM cell for mode stirring and, as a result, a device displaying the advantages of both types of chambers was obtained. Three frequency ranges were:

1. Basic frequency band of 0–150 MHz, where the TEM mode dominates. A device excited with 1 W enables linearly polarized E-fields in the calibration area at the level of 8 V/m, and estimated accuracy of the fields is within ±2 dB.
2. Transition frequency range (150–500 MHz), where the TEM mode is accompanied by a limited number of higher modes. A similar excitation allows E-fields within 8–200 V/m. In this case, polarization may be different as regards the spatial orientation of the E vector. Inaccuracy is estimated at approximately ±15 dB.
3. Stirred mode range (0.5–15 GHz), where the chamber works as a typical mode-stirred reverberation chamber. With the same excitation, E-fields up to 200 V/m can be obtained, and the inaccuracy of the field estimations is at the level of up to ±8 dB.

3.5.3   Other types of secondary standards

A primary advantage of secondary EMF standards is usually their simple design, easy application, low cost, and the ability to perform measurements under any conditions.

As a matter of fact, the role of a secondary standard may be played by any device or any source of EMF that was previously calibrated using a primary standard. As an example: It was previously mentioned that, during antenna calibration using the substitution method, the procedure could be affected by local TV or FM stations. Several years ago, a case was faced by the authors that necessitated the proposal of a method of EMF meter calibration. A local EMF meter and interference meter manufacturer arranged a test site on the roof of his building. The place was acceptable except for one factor. Use of an SRA was impossible during transmission of the local TV/FMBC station, located at a top of a hill 25 km away from the site. The problem was solved by using a selective milivoltmeter with a wideband dipole working within frequency range of 30–300 MHz and a log-periodic antenna for the range of 300–1000 MHz. The set-up was calibrated using a primary EMF standard and then used as a reference standard during calibration of a variety of meters and antennas at the site. The calibration of the reference standard was performed under conditions where no external fields could cause interference. Of course, these days, such calibrations could be more difficult because of the variety of local cellular telephony base stations and other EMF sources.

The other possibility is a direct estimation of an EMF around a radiating device. In order to illustrate this, we will present an exposure system proposed by Kuster [7]. The idea of his solution is shown in Figure 3.32. The system consists of a monopole antenna above a ground plane. In the plane around the radiator, mice are placed, radially, with their heads directed to the radiator. The system is fed from a cellular phone-like source and has been applied in biomedical experiments related to possible bioeffects in mice caused by an exposure to cellular phone radiation. The EMF distribution around the mice is a function of distance between the radiator and a point of concern. Investigations using Kuster's exposure system were applied mainly to effects on the nervous system and whether the field around the head of a mouse was of concern. It may be said that the EMF uniformity in the area of the animals heads may be assumed adequate. The field intensity is estimated on the basis of radiator excitation measurement and the geometry of the system.

Kuster's proposal ensures an exposure similar to that at a distance from a radiation source (base station, portable terminal). However, in the case of portable terminal users, apart from the absorption of EM energy radiated from the device's antenna, a head is also exposed to low-frequency (LF) magnetic field arising from Biot-Savart's law from the feeding system of the device's power amplifier. The LF H-field spectrum measured close to a Siemens C35 cellular phone is shown in Figure 3.33.

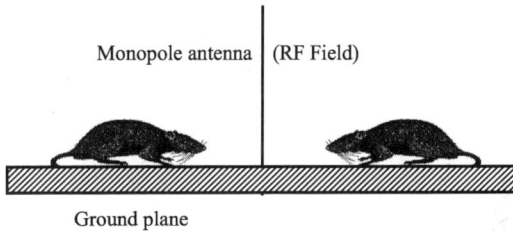

Monopole antenna | (RF Field)

Ground plane

Figure 3.32   Kuster's exposure system

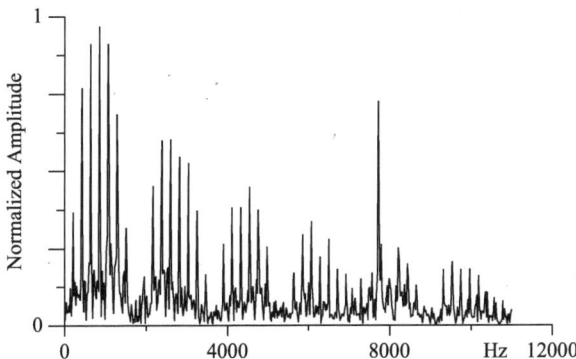

Figure 3.33   Spectrum of LF H-field radiated by a cellular phone

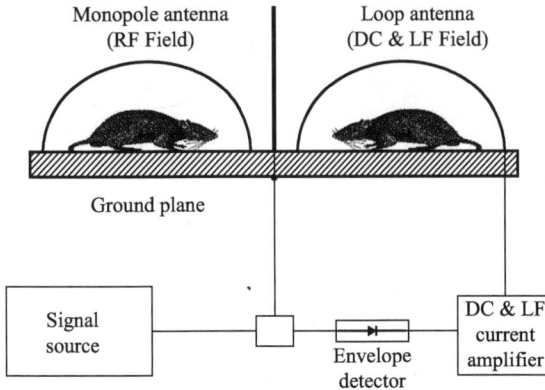

Figure 3.34   Modified Kuster's exposure system

In order to complete the exposure and bring it nearer to real conditions when portable terminals are in use, the authors proposed to complete Kuster's system with coils fed with low-frequency current of shape dependent upon the phone system investigated [45], as shown in Figure 3.34. It may be added here that the-frequency range of the current (and H-field) is similar to the brain's natural alpha and beta rhythms.

Measurements performed near portable terminals have shown that, apart from the above-mentioned EMF components, wideband noise exists, which is generated by the device's processing and controlling system in the medium-frequency (MF) range. The noise is very similar to that generated around computers and video display units. In the latter case, due to replacement of cathode ray tubes by semi-conductor displays and the consequent reduction in the radiation, the devices were found to be safe, and biomedical investigations in the area were stopped. In the case of portable terminals, a source of radiation is close to the head of its operator, and MF currents are induced in his (her) body and are measurable in any part of the body. It may also be considered here that similar frequencies are widely applied in electro- and magnetotherapy with good results, which provides further evidence of biological activity of the currents. Bioeffects laboratory studies, performed using sets similar to that in Figure 3.32, where only HF radiation is taken into account, lead to results and conclusions different from the epidemiological studies as regards the role of exposure. It may be supposed that the difference is a result of neglecting the presence of LF and MF radiations, the role of which may exceed that of HF radiation. An example of a measured MF spectrum close to a cellular phone is shown in Figure 3.35.

In order to make possible a complex exposure of an object to EMF similar to that near a portable device, a configuration is proposed similar to that presented in Figure 3.34, equipped, additionally, with a system generating EMF in the MF range. A version of the proposal is shown in Figure 3.36. HF excitation of the system is similar to that in Figure 3.34. An HF generator is applied for the purpose in order to

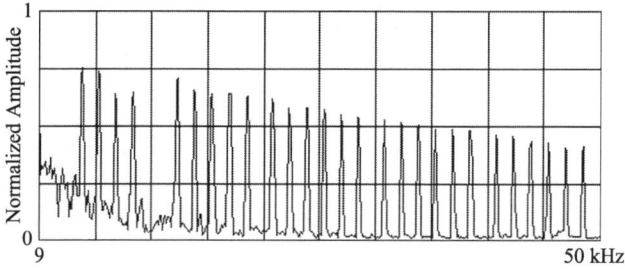

Figure 3.35   Measured MF spectrum close to a cellular phone

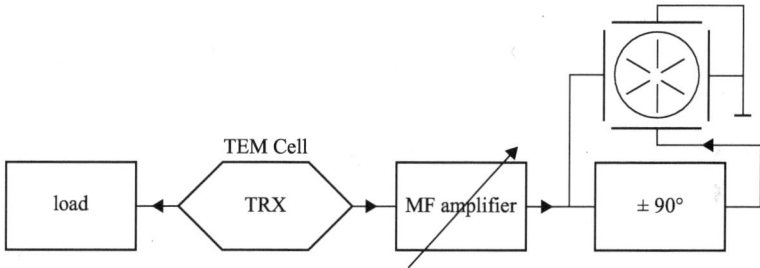

Figure 3.36   Proposed set-up for complex exposure to EMF generated by portable terminals

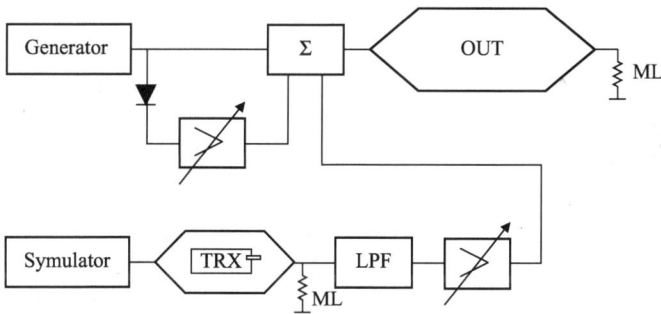

Figure 3.37   Proposed exposure system with a TEM cell

protect a portable terminal against possible destruction during long-term, full-power exposure. MF signals are picked up from an original wireless device (TRX) placed in a TEM cell. Then the signal, after amplification in a wideband MF amplifier, is fed through a ±90° phase shifter to two parallel plates, which results in generation of circulating MF field around the system. A TEM cell is required here to exclude the influence of external radiation and to ensure necessary stability. A modified version of the set presented in Figure 3.36 is shown in Figure 3.37. Here, the system of loops and electrodes was replaced by a TEM cell.

The above examples illustrate a variety of needs, and possible solutions, of different exposure systems. The freedom in their design is a very important advantage that allows the matching of a system to planned experiments. This advantage is simultaneously a disadvantage because systems designed in different centers are usually very different, which creates problems with comparison of experimental results and final conclusion formulated based on their measurements.

3.5.3.1   Transfer standards

The previously presented example with meter calibration on the roof of the manufacturer's building clearly illustrates a case where a secondary standard solved problems with calibrations. In that case, the secondary standard played the role of a transfer standard. However, its use was limited to the local conditions because its size and weight makes it difficult to move it from one place to another, which, of course, makes it impossible to use it for EMF standards comparison between labs in different countries, for example.

For international standards comparison, initiated and headed by the National Institute of Standards and Technology in Boulder, Colorado, a special, small-size, wideband EMF sensor was designed (among others). The sensor (see Figure 3.38) contains a resistive dipole loaded with a diode detector; voltage from the detector is led by a lowpass filter and a transparent transmission line to a DC voltmeter. Although the sensitivity of the device is low, and it is not free from aging effects and sensitivity to climatic conditions, it is very simple, cheap, and readily transportable.

Before the cross-comparison studies were started, a copy of the NIST transfer standard was donated to the authors for introductory preparations of comparison procedures. Another copy of the standard was circulated among several laboratories. Comparisons were made within the frequency range of 30 MHz–30 GHz. Results of the comparisons are presented in section 8.3. The standard during measurements within the anechoic chamber at the Technical University of Wroclaw is shown in Figure 3.39.

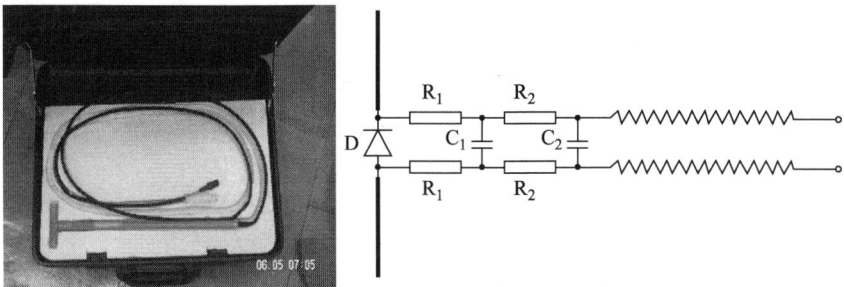

Figure 3.38   Schematic diagram (right) and a view of the NIST transfer standard in a transportation casing (left)

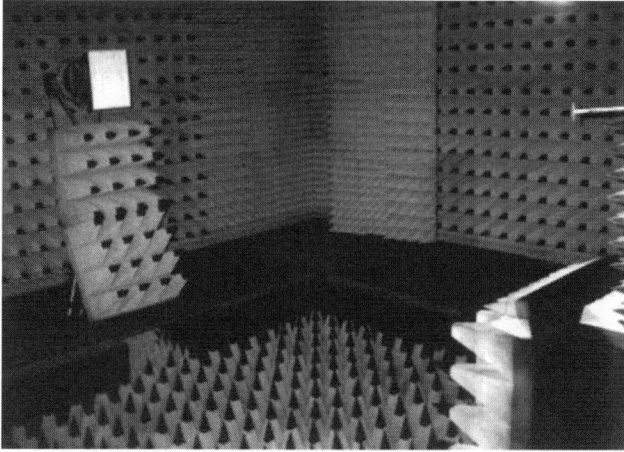

Figure 3.39 The NIST transfer standard during measurements at the Technical University of Wroclaw

3.5.3.2 Meter testers

EMF meters and probes designated for measurements related to labor safety and environmental protection are based on concepts similar to the transfer standards presented previously. They contain a field sensor (small loop or dipole antenna plus diode or thermocouple detector) connected to a DC voltmeter. EMF measurements for these purposes, especially in the near field and under conditions of many (sometimes unknown) frequency fringes, may lead to mistakes by the measuring team. Moreover, in the measurements, the source or sources of the radiation are usually unknown. Very often, these are leakages from slots in the casing of devices, insufficient screening, radiating feeders, etc. High EMF intensities may appear in absolutely unexpected places and, conversely, there may be almost no field where one is expected. In these cases, a natural first conclusion is that the meter is not working correctly, not to mention a possibility of damage to a meter during transportation on trackless roads and/or of effects on its performance due to atmospheric conditions and the local environment. In order to make it possible to check the meters, the authors designed several types of EMF meter and probe testers. The oldest types were designed only for meters used by Polish control authorities and manufactured in Poland. The newest design makes it possible to check almost any type of meter designated for EMF surveying available on the market. A front panel of the old type of MT USMEH-1, during testing an E-field probe, is shown in Figure 3.40, while a newer one is shown in Figure 3.41. The latter contains four generators, which makes it possible to check E- and H-field probes as well as probes for power density measurements from low frequencies up to microwave frequencies. A probe under test is placed into a "pocket," around which are coils and plates generating the required field component in the required frequency range. The newer type allows checking directional pattern as well by turning the probe in

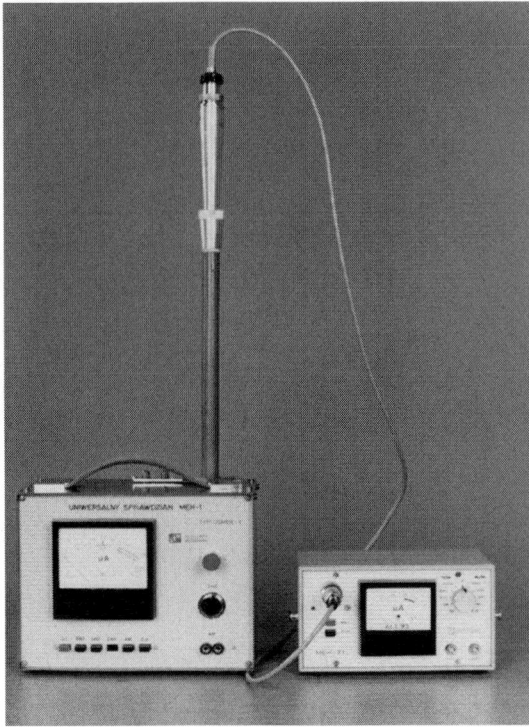

Figure 3.40   An E-field probe testing with USMEH-1 type tester

the pocket. The older type of tester allowed regulation of the field level. In the newer one, the field intensity is constant and is stabilized for any frequency/component combination. Testing procedures for different types of meters are given by the manufacturer of the device. Unknown meters may be tested, with the use of the MT, by way of comparison of the tested meter's indications when calibrated in laboratory conditions, before field measurements and during them. The procedure is equivalent to introductory MT calibration with the use of a meter, followed by meter testing using a MT calibrated in this way.

The testers illustrated above are excited from CW sources. It is a rule that all types of meters are calibrated in CW fields, which assures a universal standard of calibration. However, more and more sources work with different types of pulsed modulations, which is discussed in section 7.4. Here, an MT is proposed that assures the ability to test a meter in a CW field, a pulse-modulated field, and in trains of pulses. A schematic diagram of the device is shown in Figure 3.42. The device includes three generators. When power is fed to the CW generator, the device ensures a CW field at a selected frequency; in the case of feeding the rectangular pulse generator 1 (PG_1), the device generates a pulse-modulated field, and the duration time of the pulses and their frequency may be selected in required ranges. In the third case, when power is fed to the second rectangular pulse

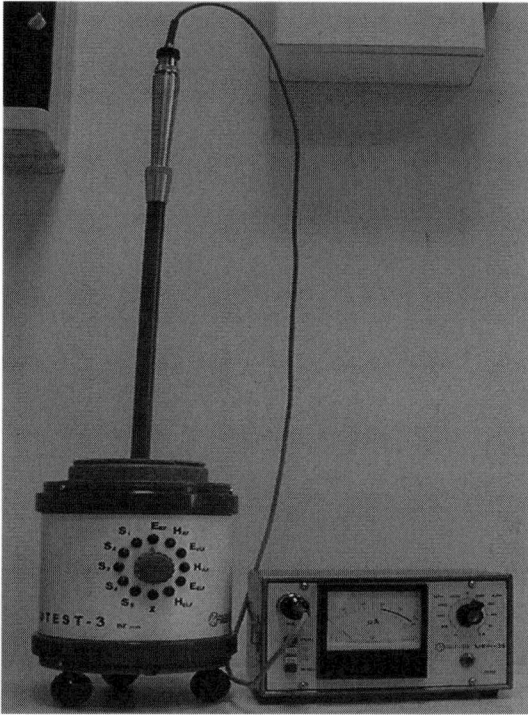

Figure 3.41 Calibration of a meter with a newer type of MT

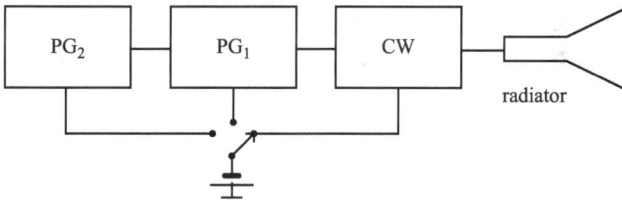

Figure 3.42 Block diagram of a universal MT

generator (PG$_2$), the device generates trains of pulses, equivalent, for instance, to nonstationary EMF near a rotating radar antenna. Again, the duration and frequency may be chosen in accordance with metrological requirements. The concept is identical to that usually applied in radar stations; however, the pulse modulation may be doubled in order to have continuous pulse modulation (typical of radars) and trains of pulses (as EMF observed at a point away from a rotating radar antenna). The device excites a radiator (an example with a horn antenna is shown in Figure 3.42), and the testing procedure is similar to that presented above.

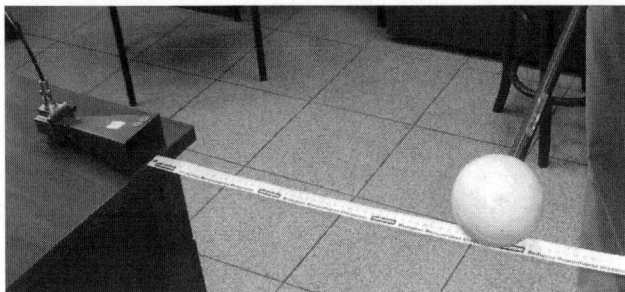

Figure 3.43   A practical design for a universal MT

A practical solution is shown in Figure 3.43. A typical Gunn diode microwave generator is loaded with a horn antenna. The generator is fed from a set of two generators, as shown in Figure 3.42. It allows a CW field generation, pulse modulation, and trains of pulses, typical of nonstationary EMF. At a distance is placed a tested probe. Indications of the probe should be similar during primary testing under laboratory conditions and during control testing in a place of measurement.

Chapter 4

Accuracy analysis of EMF standards with dipole antennas

4.1 Accuracy analysis

Before we start to undertake an accuracy analysis of standard EMF generation and measurements, as presented in previous chapters, we should briefly discuss problems related to measurement accuracy and a classification of errors that could be faced during the measurements.

Any magnitude or phenomenon may be recognized as "known" when we are able to describe it in appropriate values, i.e., when we can measure it with known (required) accuracy. Development in any area of science and technology is accompanied by a development of metrological techniques specific to that area. It is evident that there is not a single ideal metrological methodology, condition or tool; every measured parameter (magnitude, phenomenon) is taken with finite accuracy, every one is loaded by an error. Every measurement is accompanied by a certain uncertainty (confidence) level, specific to any metrological procedure and metrological tools used. Any measured value is given within an uncertainty range; if the results are to be recognized as correct, the true value of the measured magnitude should be within the range as well. During measurements, especially when using indirect methods (as in the case of EMF metrology), we have different sources of error that disturb the measurements. Because of this, it is of primary importance to be able to identify particular reasons for the errors, to understand the types of error and the sources of error in any considered case. Identification of the errors, their mutual correlation and essence, and, finally, the possibility of estimating the total error of an applied procedure may lead to elimination of particular errors or, at least, limiting their role, as already mentioned in the introduction.

This chapter and the following ones are devoted to the problems of error estimation in measurements for the standard EMF generation and measurement. We show the complexity of the problems as well as sources of error identification and values of error estimations. This understanding leads to the possibility of estimating the uncertainty of a procedure and a class of standards considered.

Every result of a measurement is subject to errors introduced by the limited accuracy of the procedure (method) applied, inaccuracy in the meters and other auxiliary equipment, imperfection of humans' minds and the possibility of mistakes, as well as varying conditions and circumstances during the measurements, etc.

All these factors affect the results of the measurements and, as a result, there appears a difference between the measured value and the true value of a magnitude that always remains unknown or can be known with a certain accuracy. The metrological experience of the authors, as well as literature studies, show that the problem of accuracy is often misunderstood, and this degrades the accuracy of many procedures and/or leads to a false interpretation of measurements (experiments). The problem is especially important in the case of electromagnetics because of the variety of indirect procedures and measurements that lead to the final estimation of the required field component and its parameters on one hand, and the people performing the measurements, who may represent fields other than electromagnetics (metrology, power engineering, biology, medicine) and who are often less familiar with electromagnetic phenomena, on the other. An example of the latter case is results presented in the literature on indoor EMF measurements with the use of resonant and directional antennas or E-field measurements under similar conditions as when using meters equipped with loop antennas.

As has been said, EMF is not measurable directly. The standard EMF generation or measurement includes several steps between a generator and a field in the STA method or between the field and a voltmeter (indicator) in the SRA method. Every one of the steps is loaded with errors specific to it. In previous chapters, only problems with EMF homogeneity were outlined in order to show the necessity of the choice of an appropriate method for standard field generation as well as to show what level of errors might be expected. Problems with the method, geometry of propagation, excitation measurements, and other considerations leads to the fact that the standard EM field is one of the least accurate among standards of all physical magnitudes. If we would like to estimate the total inaccuracy of a standard, we have to take into account all factors that could disturb selected procedures and that limit accuracy of the standard. Such a procedure should be applied individually for any particular case or solution. Here, it is impossible to generalize and, as a result, estimations presented are only of an illustrative character. However, the classification of errors, presented below, and their analysis may be helpful in understanding the problem.

In the theory of measurements, sources of error are divided into three groups:

1. random errors
2. systematic errors
3. gross errors (mistakes)

Random errors, in the case of EMF standards, may be caused by fluctuations of internal and external conditions, i.e., instability of applied equipment, variations of weather conditions, illumination, etc. Their precise analysis is almost impossible, but their presence is apparent when a series of measurements are repeated several times for the same configuration but the results of the measurements are different. In the case of EMF measurements taken at an OATS, the influence of external EMI may be observed that is impossible to take into consideration. The EMI may come from local or even distant sources of radiation, airplanes, cellular phones, radiotelephones, and others. They are most noticeable when low-level EMFs are in use.

An example of an absolutely unexpected, incidental error that surprised the authors during their first dipole antenna calibration at an OATS using the SRA method was that a source of remarkable differences in repeated calibrations was found in the sunlight and temporal variations of its intensity on the non-light-protected detection diode in the SRA.

Because of the random character of the factors mentioned and the measuring errors caused by them, their total elimination is impossible, as is the introduction of any correction factors to compensate for them. The only possibility of detection and estimation lies in repetition of the measurements and in taking a statistical approach to the results obtained.

As a result of a series of measurements, we may have several results that differ from each other according to a standard distribution that may be characterized as follows:

- The errors are subject to continuous random change.
- Every individual result of measurement is characterized by one error.
- Small errors are more probable than big ones.
- Positive and negative errors are equally probable.
- The most probable error is equal to zero.

If a searched value, y, is a function of n mutually independent measured values $x_1, x_2, \ldots, x_n$, i.e.:

$$y = f(x_1, x_2, \ldots, x_n) \tag{4.1}$$

then a calculation of a variation of the value y (dy) caused by small variations of values x_i (dx_i) is given by a differential:

$$dy = \sum_{i=1}^{n} \frac{\delta y}{\delta x_i} dx_i \tag{4.2}$$

Equation (4.1) illustrates the absolute error of measured magnitude y. As an example, let's take into account (3.11):

$$E = \frac{V}{D} \tag{4.2a}$$

the absolute error dE is:

$$dE = \frac{1}{D} dV + \frac{V}{D^2} dD \tag{4.2b}$$

Usually, it is the relative error, i.e., dy/y, that is of concern. In our example, we may write:

$$\frac{dE}{E} = \frac{dV}{V} + \frac{dD}{D} = \delta_V + \delta_D = \delta_E \tag{4.2c}$$

where: δ_E, δ_V, and δ_D are errors of E-field intensity, voltage, and distance measurement, respectively.

The above analysis shows a maximal value of the measuring error that is an algebraic sum of separate errors; i.e., it assumes that during the measurements there is a probability that the separate errors may take their maximal values. In metrological practice is assumed that this approach is improbable and a variety of statistical approaches are offered. The most popular is based upon a normal probability distribution, i.e., the square law of error distribution (Gauss's law of error distributions). Taking this law into account and changing infinitely small variations dx_i to finite ones Δx_i and assuming that the error may be positive or negative, (4.2) may be rewritten in the form:

$$\Delta y = \pm \sqrt{\sum_{i=1}^{n} \left(\frac{\partial y}{\partial x_i} \Delta x_i \right)^2} \tag{4.3}$$

The relative error $\Delta y / y$ is then:

$$\delta_y = \pm \sqrt{\sum_{i=1}^{n} \left(\frac{\partial y}{\partial x_i} \delta x_i \right)^2} \tag{4.4}$$

Or, in our example:

$$\delta_E = \pm \sqrt{\delta_V^2 + \delta_D^2} \tag{4.4a}$$

The error in percent is obtained by multiplication of the above by 100%.

Although there is no method that would allow an elimination of random errors, they could be partially limited by increasing the number of measurements as well as improving the metrological techniques, e.g., automatization, CAD, etc. However, every increase in the number of repeated measurements at the same point suggests an optimization procedure that would allow a reduction in the number of measurements and a simplification of the measuring procedure, which would also make it more stable to external measurement conditions. In the case of repeated measurements, in (4.1) and subsequent equations, a mean value of x_i should be applied, instead of x_i, but the dx_i (or Δx_i) remains unchanged because it is estimated (evaluated) in a separate way. In general, it may be supposed that as $n \rightarrow \infty$, y approaches the correct value of measured magnitude.

The next type of errors are systematic ones. Systematic errors are errors of constant values or values alternating in accordance with a rule. These errors may be caused by uncertainty of applied meters and other auxiliary equipment as a result of imperfections in their design, inaccurate calibration, application in conditions other than nominal ones (e.g., a mismatching), etc. In order to find these errors, it may be necessary to undertake comparative measurements using similar devices or methods, e.g., repeat EMF measurements (calibrations) on an OATS, anechoic chamber, TEM cell, using a transfer standard, or others. Such a procedure creates a possibility of identifying systematic errors (and the causes of errors) and taking them into account, and, as a result, limiting or eliminating them. Such an approach is used, for instance, when an EMF probe is calibrated in a TEM cell (Chapter 6). Probe sensitivity changes due to a change in its antenna input impedance caused by the presence of reflections in the cell walls may be corrected and, as a result, referred to the free-space conditions, by use of appropriate correction factors that may be found in both theoretical and experimental ways. Other examples of systematic errors are: possible variations in measuring conditions or temperature variations during measurements, the presence of mutual couplings between a standard and an OUT, EMF deformations caused by the OUT, and others. The presence of these factors requires continuous control of all of them when measurements (or calibrations) are performed.

While random errors may be positive or negative during a series of measurements, systematic errors are, as a rule, of a permanent character. They can be negative or positive ones, but the character is invariant during measurements performed in unchanged conditions.

The last kind of errors that may occur in any kind of experimental work are gross errors, which are of human origin rather than arising from appropriate use of methodology or measuring equipment. They can be a result, for instance, from:

- a mistake in measurement notation or readings from an improper scale of an indicator or an improper indicator,
- introduction to a measuring set of changes that may cause the presence of systematic errors that were previously neglected, for instance, a change of voltmeter, power meter, or other device whose impedance is different from that of the devices previously applied and, as a result, introducing mismatches into the system,
- unintentional change of measuring conditions, for instance, distance between antennas, or inappropriate substitution, as when a standard antenna is replaced by a calibrated one,
- unnoticed, usually short-term, changes of measuring conditions caused, for instance, by the presence of external EMI, alternations of power line voltage, flying birds during OATS experiments, etc.,
- a fault or damage to equipment due to overloading (a thermocouple, diode detector, matched load) or other causes of electric or/and mechanic damage,
- measurements performed by inexperienced people or a change of measuring team during measurements and application of a different approach (interpretation) to measuring procedures and their results.

Gross errors or mistakes may be a result of many causes, and it is not possible here to analyze them in a systematic way, as in the case of random and systematic errors. Their presence only illustrates the necessity of focusing the attention of measurement personnel on the work they do. Very helpful here is the experience of people and the ability to check the measuring set-up. A warning of the presence of a disorder or an unexpected error may be the difference between results obtained so far and the next ones, or a notable difference between measured values and calculated ones (or those estimated in another way). When the results of measurements are analyzed, results which differ considerably in relation to the others, "outliers," may be neglected.

When an analysis of accuracy of measurements (calibrations) is performed, all sources of error, as previously discussed, should be taken into consideration. The above short and simplified presentation of the problem is included only in order to focus attention upon the role of error, especially when standards are discussed. EMF standards are here "privileged" due to a number of factors that do not exist in the measurement and standardization of other physical magnitudes. These are among the least accurate of standards. Detailed analysis of the problem may be found in many studies, for instance, in [NIST 1297].

4.2   Choice of the OATS

EMF standards with dipole antennas are usually applied within the frequency range 30–1000 MHz. Because of quite large sizes of the standard, needed especially below 100 MHz, as well as above when larger antennas are calibrated, and the costs of large anechoic chambers, calibrations are often done at open area test sites. At higher frequencies, any calibrations are mostly performed in an anechoic chamber. The advantages of the latter include better accuracy of the procedures (no couplings with external EM environment, taking into account direct rays only) and the ability to perform measurements irrespective of weather conditions. The main disadvantage of the anechoic solution is its cost. Where expenditures for chamber construction are not acceptable, or calibrations are performed infrequently, the OATS is an almost ideal solution. We will briefly discuss problems related to the choice of appropriate terrain and criteria permitting an evaluation of terrain applicable as an OATS.

First, the chosen terrain should be at a considerable distance from the nearest sources of EM radiation, and preferably free of strong EM fields from external sources. Nowadays this condition is not easy to fulfill, even on a global scale. Local, or even distant, TV and BC stations, cellular phones, radars, and a variety of different radiating devices and systems that create E-fields at levels of 0.1–0.2 V/m are almost everywhere; this is the reason for the old NIST proposal to use natural caves or old mines for calibrations purposes.

An acceptable external EMI level depends upon the calibration method to be used and the types of antennas (devices) that will be calibrated. The environment is almost irrelevant when the STA method is to be used for the calibration of antennas loaded with selective devices. On the other hand, the most sensitive to the external environment is the SRA method and wideband device calibration. In any case, checking the local EM environment and its variations before the measurements start is advised.

The terrain should be free of any objects that could cause reflections. Overhead power lines, telephone cables, buildings, metallic towers, and even metallic fencing should be farther from the OATS than 10 × the planned greatest distance between the transmitting antenna and the receiving one. The terrain should be electrically

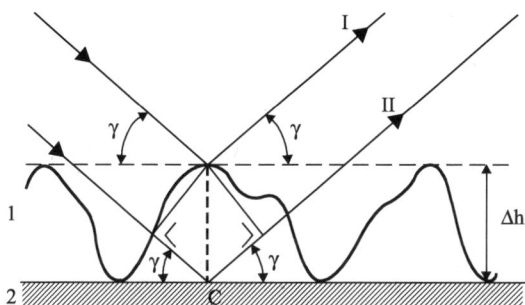

Figure 4.1 A plane wave reflection from an irregular surface

flat and homogeneous. When both the antennas are elevated, the resultant field at the place of the receiving antenna is a result of interference of a direct ray and a reflected one. An area essential for the reflected ray to form is equal to half of the first Fresnel zone. Irregularity of the ground in the zone or, more strictly, inhomogeneities of the ground structure, and its electric parameters to a depth equal to the EM wave penetration depth at the frequency of concern should not cause phase errors of the reflected wave exceeding acceptable levels. The problem may be considered on the grounds of the Rayleigh criterion.

If, at an irregular surface of two layered media, a plane EM wave is incident at angle γ, and the surface is assumed to be a lossy layer (1) covering a perfectly conducting layer (2) (Figure 4.1), then the difference in length between ray I (reflected form the surface of layer 1) and II (reflected from the surface of the layer 2) is $2\Delta h \sin \gamma$. This causes a phase difference $\Delta \varphi$ between the rays of:

$$\Delta \varphi = \frac{2\pi}{\lambda_m} 2\Delta h \sin \gamma \qquad (4.5)$$

where: Δh – thickness of the first layer,

λ_m – wavelength in medium 1:

$$\lambda_m \cong \frac{\lambda}{\sqrt{\varepsilon_r}} \qquad (4.6)$$

ε_r – relative permittivity of layer 1.

If we accept that the maximal phase difference of the rays creating the phase front of the reflected wave is $\Delta \varphi = \pi/4$, then the maximal allowable thickness of the first layer Δh_{max} is:

$$\Delta h_{max} \leq \frac{\lambda_m}{16 \sin \gamma} \qquad (4.7)$$

Figure 4.2 shows the results of calculations of h as a function of γ for the assumption that $\lambda = \lambda_m$ [4]. The figure allows an introductory estimation of a terrain's applicability for calibrations. However, it is to be remembered that, as a rule,

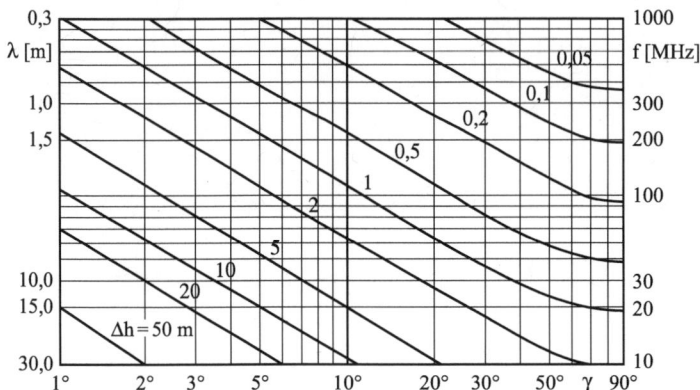

Figure 4.2   Acceptable thickness of the first layer Δh_{max} as a function of γ and λ

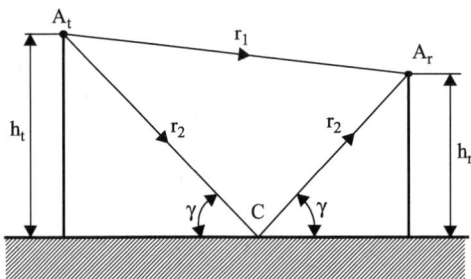

Figure 4.3   Two horizontally polarized dipoles over a ground plane (C indicates a reflection point)

$\lambda_m < \lambda$ and, as a result, the value of Δh_{max} read from the figure should be reduced. The figure illustrates numerically the evident conclusion that a more accurate selection of the OATS is required for higher frequencies. The above discussion presents the problem of site selection when a natural surface is planned to be accepted. In the case of covering the surface with a conducting mesh (reference ground), this problem may be neglected, provided that the mesh is fine enough that it appears as a continuous surface at the maximum frequency of interest.

The above criterion concerning irregularities of the terrain at the OATS should be fulfilled in the area of the first half of the Fresnel zone. If a transmitting antenna A_t is elevated at h_t and a receiving one A_r at h_r (as shown in Figure 4.3), then the criterion should be fulfilled at a plane of an ellipse whose axes are given by (see Figure 4.4):

$$a = \sqrt{\frac{\lambda\, h_t\, h_r}{(h_h + h_r)\sin\gamma}} \tag{4.8}$$

$$b = a\sin\gamma \tag{4.8}$$

where: a and b – half axes of the ellipse.

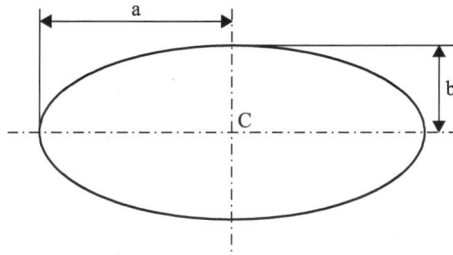

Figure 4.4   Sizes of the first Fresnel zone

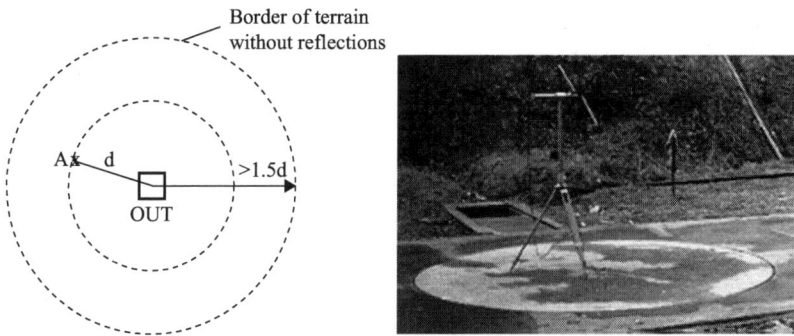

Figure 4.5   Recommendations related to requirements of OUT sizes free of reflections and designated for measurements and calibrations of quite large objects placed directly at the ground (left), and a practical illustration (right)

In order to unify the criteria of the OATS selection that would make possible the national and/or international comparison of the results of measurements performed in different centers, many studies were carried out under the auspices of the International Special Committee on Radio Interference (CISPR: Comité International Spécial des Perturbations Radioélectriques), a special committee under the sponsorship of the International Electrotechnical Commission (IEC). The studies were concerned with specific measurement methods for different purposes. Ground recommendations were formulated for EMC-related studies within the frequency range 10 kHz to 18 GHz (CISPR Publication No. 16). Recommendations related to requirements for OATS sizes without reflections and designated for measurements and calibration of large objects placed directly at the ground are presented in Figure 4.5. The requirements for small-size objects, placed on a support at 0.8 m above the ground surface, are presented in Figure 4.6. Figure 4.7 presents minimal sizes for the OATS. For standard distances between a transmitting antenna and an OUT, $d = 3$, 10, or 30 m. The minimal diameter of the circle at which the OUT is placed should be $d_2 = d_1 + 2$ m, and the minimal space around the antenna $W = D + 1$ m.

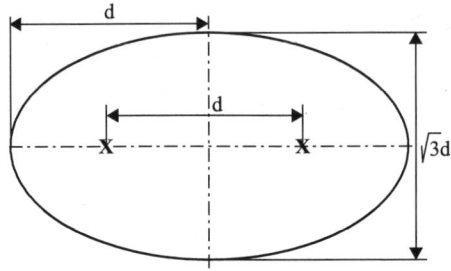

Figure 4.6   Requirements for small-size objects, placed on a support at 0.8 m above the ground surface

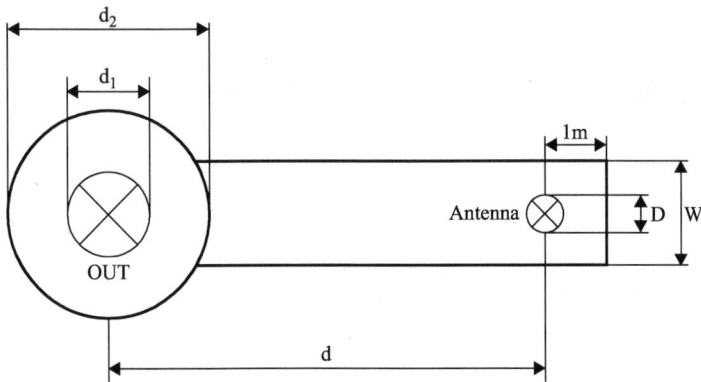

Figure 4.7   Minimal sizes of the OUT

4.3   Measurement of the electrical parameters of the ground

As previously mentioned, even without regard to previously performed measurements of the ground parameters at the OATS, it is suggested that the measurements be repeated before any new series of calibrations. The parameters measured may affect the reflection coefficient as well as the input impedance of antennas used. Although it is possible to use quick electrode methods for parameter measurements, they provide accurate measurements of the parameters of the surface layer, whereas the reflection is formed at a depth equal to the penetration depth, and this depth is a function of frequency. Thus, while an accurate measurement of the ground parameters, averaged over the penetration depth, is of concern, measurement of the parameters at a given frequency is necessary.

The most frequently applied methods are based upon the ellipse of polarization measurement, measurement of an input impedance of an antenna at a distance from the ground, and a reflection factor measurement. The first method is based upon a polarization measurement of a ground wave from distant BC stations and is usually applied at medium and long waves. Measurements of input impedance are inaccurate and may be affected by local inhomogenities of the ground. The most accurate, although the most involved, is the reflection factor measurement. The method is briefly discussed below.

The idea of the method is shown in Figure 4.8 [2]. Above a ground of concern, at height h_t, a transmitting antenna A_t is placed. Below it, at elevation h_r, is placed a receiving antenna A_r. The two antennas are placed in a common plane, parallel to the earth's surface and parallel to each other and to the surface.

Due to reflections from the earth's surface, a standing wave will appear above the ground. Moving the receiving antenna up and down, a near maximum and minimum of the standing wave should be found. Then the value of the reflection factor Γ_h, for the horizontal polarization, may be calculated using the formula

$$|\Gamma_h| = \frac{\left|\dfrac{V_{r\,max}}{V_{r\,min}}\right| \cdot (1+\delta) \cdot (h_t - h_{r\,min})^{-1} - (h_t - h_{r\,max})^{-1}}{\left|\dfrac{V_{r\,max}}{V_{r\,min}}\right| \cdot (1+\delta) \cdot (h_t + h_{r\,min})^{-1} + (h_t + h_{r\,max})^{-1}} \qquad (4.9)$$

where: V_{min} and V_{max} – voltages measured at the receiving antenna's input for $h_r = h_{rmax}$ and $h_r = h_{r\,min}$, respectively,

$h_{r\,min}$ and $h_{r\,max}$ – elevations of the receiving antenna corresponding to the nearest minimum and maximum of the standing waves,

δ – estimated systematic error of the procedure.

The phase angle of Γ_h is:

$$\Theta_h = 2\left[\frac{(2n-1)\pi}{2} - \frac{2\pi\,h_{r\,min}}{\lambda_0}\right] \qquad (4.10)$$

or

$$\Theta_h = 2\left[\frac{2\pi\,h_{r\,max}}{\lambda_0} - (n-1)\right] \qquad (4.11)$$

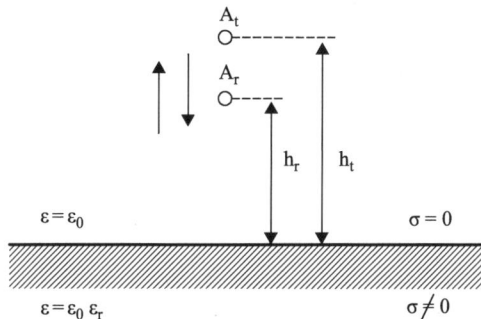

Figure 4.8   Positioning of antennas for reflection factor measurement

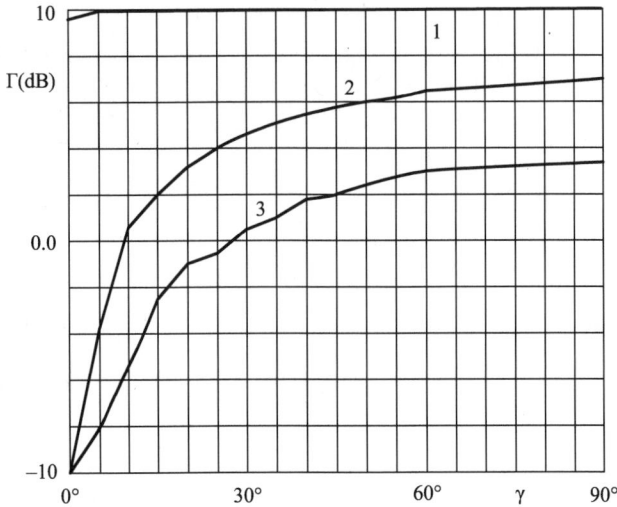

Figure 4.9   Reflection factors calculated for (1) aluminum plate, (2) dry sand, (3) asphalt [48]

where: n = 1, 2, 3, ... – a number of the minimum or maximum counted from the ground surface.

Figure 4.9 presents results of calculations of the reflection factor for different grounds, as a function of angle of incidence γ [48].

Methods for measurement of the ground parameters are rather difficult and time consuming. Moreover, the measurements should be repeated before any series of measurements (due to possible weather influences). It is evident that the problem may be neglected if the ground is covered by a conducting mesh.

4.4   Analysis of the accuracy of the SRA standard

If a dipole antenna is illuminated by a homogeneous EMF and its arms are parallel to the E-field vector, then the electromotive force e_A induced by the field in the antenna is:

$$e_A = E\, h_{eff} \tag{4.12}$$

where: h_{eff} – effective length of the antenna.

The effective length of a thin, symmetric dipole antenna (as shown in Figure 4.10) is given by:

$$h_{eff} = \frac{2\,[J_0(kh) - \cos kh]}{k \sin kh} \tag{4.13}$$

where: J_0 – Bessel function of first kind and zeroth order,
       k – propagation constant.

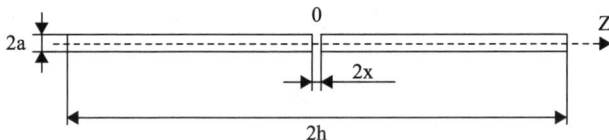

Figure 4.10   A symmetric dipole antenna

The formula is valid for an assumed sinusoidal current distribution in the antenna and a → 0 and x → 0. As a rule the length of the SRA is $2h = \lambda/2$. Effective height of the half wave dipole is $h_{eff} = \lambda/\pi$. The calculation is fully correct if $2h/a \rightarrow \infty$. For antennas of finite slenderness, the current distribution along the antenna is a function of its electric length 2 kh, and the slenderness ratio and distribution are expressed by:

$$\frac{I_z}{I_0} = f(kh, K_0) \qquad (4.14)$$

where: I_0 and I_z – current at the input of the antenna and at distance z from its center,

K_0 – slenderness ratio:

$$K_0 = 120 \left(\ln\frac{2h}{a} - 1 \right) \qquad (4.15)$$

and the effective length is calculated for real current distribution given by (4.14) as:

$$h_{eff} = \int_{-h}^{+h} \frac{I_z}{I_0} dz \qquad (4.16)$$

The formula makes possible a calculation of the h_{eff} of any dipole, leading to the following expression:

$$h_{eff} = \frac{\lambda}{4} \left(1 - \frac{0.2257}{\ln\dfrac{\lambda}{2a} - 1} \right) \qquad (4.17)$$

To illustrate the increase of the effective height of a cylindrical dipole in relation to the half wave dipole, Figure 4.11 shows results of calculations of the percentage increase of its length as a function of K_0 [16].

Figure 4.12 presents the results of calculations of size variations in a cylindrical symmetrical dipole, in relation to a half wave dipole, in order to tune it to resonance. This is to focus attention on the fact that the presented estimations of the

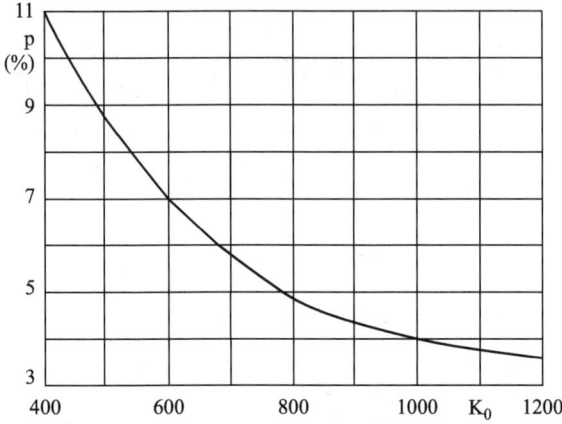

Figure 4.11   *Percentage increase of the effective length of a cylindrical half wave dipole in relation to λ/π as a function of K_0*

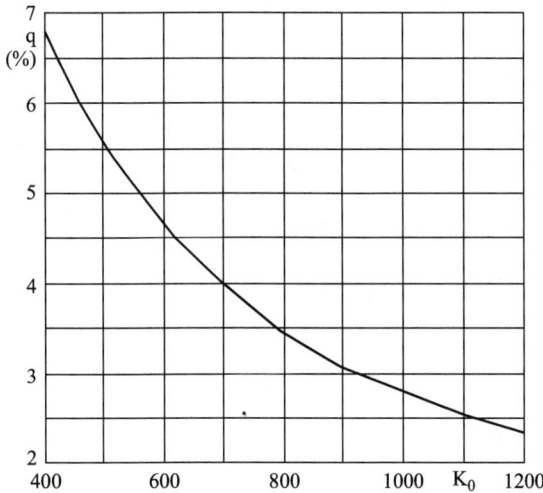

Figure 4.12   *Percentage reduction of the length of a cylindrical dipole in relation to half wave dipole*

cylindrical dipole's length is especially important at the highest frequencies, where, for mechanical reasons, it is impossible to construct dipoles thin enough to use simplified formulas and presented correction factors are to be used.

At the input of the SRA, the electromotive force e_A or voltage V_1 is measured if the antenna is loaded by a Z_1; according to the Thévenin's theorem we have:

$$V_1 = e_A \frac{Z_1}{Z_A + Z_1} \tag{4.18}$$

The input impedance of the antenna is a function of many variables:

$$Z_A = f(h, h_r, \Gamma_h, K_0) \qquad (4.19)$$

In order to be able to neglect necessary measurements of the ground conductivity (or if a mesh is in use), estimation of the antenna input impedance corrections due to the presence a conducting medium near it, as well as other sources of error, the electromotive force at the antenna inputs should be measured. However, this case requires that the condition $Z_a \ll Z_l$ be fulfilled.

The input impedance of a half wave symmetric dipole antenna is $Z_A \approx 70\ \Omega$. Thus, it is enough if the loading impedance $Z_l > 1\ \text{k}\Omega$. There is no technical problem to designing a symmetric HF voltmeter whose input impedance would fulfill this requirement. The main disadvantage of the solution is wideband sensitivity of the set for any external EMI. Nevertheless, we will discuss three possibilities for e_A measurement.

4.4.1 *Electromotive force measurement with a thermocouple*

When emf at the input of the SRA is measured with the use of a thermocouple, the E-field intensity is given by:

$$E = \frac{\alpha\, I}{h_{eff}} (R_A + R_{th}) \qquad (4.20)$$

where: I – current of the thermocouple,

α – a correction factor that represents the thermocouple characteristics,

R_{th} – resistance of the thermocouple,

R_A – input resistance of the antenna (it is assumed that the antenna impedance at resonance is only real).

The mean square error of the E-field assignment, measured with the use of a thermocouple (δ_E), as given by (4.20), is:

$$\delta_E = \sqrt{\delta_1^2 + \delta_2^2 + \left(\frac{R_A}{R_A + R_{th}}\right)^2 \delta_3^2 + \left(\frac{R_{th}}{R_A + R_{th}}\right)^2 \delta_4^2} \qquad (4.21)$$

where: δ_1 – error of the current measurement,

δ_2 – error of the h_{eff} of the SRA estimation,

δ_3 – error of the R_A estimation (measurement),

δ_4 – error of the R_{th} measurement.

Error δ_1 is an error of the thermocouple calibration, usually at DC; it is assumed that it is equal to the current measurement error; its value can be expected to be approximately $\pm 2\%$.

Error δ_2 includes inaccuracy of the geometrical sizes of SRA measurement, inaccurate estimations of the antenna length correction factor due to its finite slenderness and nonsinusoidal current distribution. The error value is estimated as not exceeding $\pm 2\%$.

The input resistance of an antenna is a function of its slenderness, electrical length, elevation above a ground (or presence in its neighborhood of any conducting objects), and electrical parameters of the ground. Measurements of the antenna's input resistance may be done, under the worst conditions, with inaccuracy $\delta_3 < 2\%$.

Error δ_4 at frequencies around 300 MHz should not exceed $\pm 2\%$; thus, the E-field measuring error, when a thermocouple is applied for the emf measurement at the input of a SRA and for $R_A = R_{th}$, may be estimated to be a little greater than $\pm 3\%$. Of course, the above estimation is an example only and a similar procedure should be applied to any standard set, frequency, resistance of thermocouple applied, and other factors that would be able to decrease the accuracy of a calibration.

The factors discussed above, limiting the accuracy of the standard, only take into account the SRA and its load under static conditions. When the total accuracy of the standard is estimated, the approach should take into account possible variations of factors in space, time, and the frequency range as well as factors associated with external conditions, including:

1. A systematic error caused by the frequency dependence of the thermocouple parameters due to its efficiency decreasing as the frequency increases, and possible resonant phenomena. It may be noticed that the error δ_1 is frequency dependent.

2. A systematic error caused by temperature variations during measurements. For example, a thermocouple is calibrated at a given temperature but its parameters change as the temperature changes. This is particularly important when measurements are performed at an OATS where there is no protection against weather condition changes. It is estimated that the parameters of the thermocouple may change by about 1%/10°C. Both of these errors are "systematic," as it is possible to continuously control them and take appropriate corrections into account.

3. An error caused by inaccurate substitution of an OUT in place of the SRA. This error should not exceed $\pm 1\%$.

4. An error resulting from the SRA input impedance changing due to the presence of a conducting ground. This phenomenon is of concern when voltage or current in the SRA is measured. To illustrate the scale of this phenomenon, Figure 4.13 shows the influence of the elevation of the SRA above a ground where the reflection factor is $\Gamma_h = 0.8$ and for three values of R_{th} as a function of h_r/λ. The error results from the application of Thévenin's theorem (4.18). It may be noticed from the figure that the most favorable is where the elevation of the SRA (h_r) is equal to [3]:

$$h_r = n \frac{\lambda}{4} \qquad (4.22)$$

The error may be easily eliminated by the way of measurement of the SRA input resistance R_A during each step of the calibrating procedure (any antenna

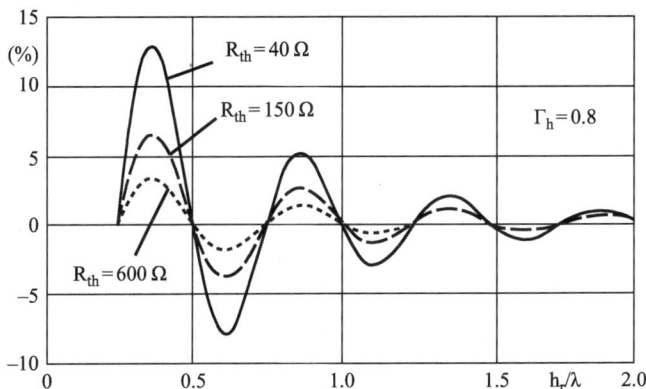

Figure 4.13 E-field calibration inaccuracy due to presence of real earth and different R_{th} [3]

configuration, any frequency). The same procedure may be repeated when the SRA is replaced by a calibrated antenna. Total inaccuracy of the dipole antenna calibration at an OATS using the SRA method and a thermocouple measurement of emf at the antenna input may be estimated at a level not exceeding ±4% at frequencies below 300 MHz and ±5% up to 1 GHz.

4.4.2 Voltage measurement using a selective voltmeter

The method of the emf measurement discussed in the previous section, because of its wideband nature, may be susceptible to external interference; making it impossible to perform calibrations, especially at an OATS. This disadvantage may be fully eliminated if voltage at the input of an SRA is measured using a selective voltmeter or spectrum analyzer, as shown in Figure 4.14.

This set-up is identical to calibration using the SRA method and differs only in the use of a selective microvoltmeter for the output voltage of the SRA measurement. Accuracy estimations of the procedure are very similar to that presented above. In contrast to the above, it requires good matching of the SRA to its load (Z_l), i.e., $Z_A = Z_l$. In resonant conditions, we may assume that $Z_A = R_A$, while Z_l is a standard wave impedance of a microvoltmeter and interconnecting

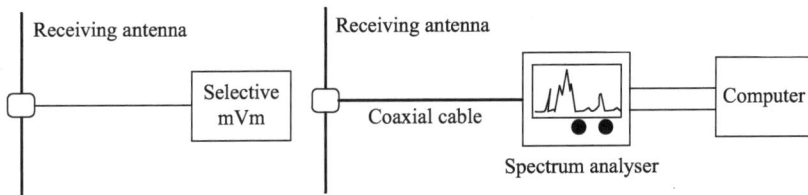

Figure 4.14 Calibration with the use of a selective voltmeter (left) and a spectrum analyzer (right)

cables. Due to the matching conditions, we may assume that the voltage at the SRA input (V_A) equals to $e_A/2$. However, the voltage measured by the microvoltmeter is reduced by the attenuation of the connecting cables, and the attenuation has to be taken into account in calculations of the final results of calibrations. Similar to the previous approach, this is sensitive to the presence of a conducting ground (as shown in Figure 4.13), and, similarly, elevation of the SRA is recommended. Disadvantages of the method include the cable attenuation previously mentioned, the possible role of the cable as a reflecting object, and technical problems with antenna matching networks, including their losses. Matching problems are especially troublesome when calibrations are planned over a wide frequency range. The disadvantages lead to limited use of this approach in the role of primary standards. In a modified version of the method, a wideband antenna is applied using the SRA instead. This allows wideband, automatized measurements; however, in order for a set to be representative of a typical example of a secondary standard, it requires an introductory calibration.

4.4.3   Electromotive force measurement using a diode detector

In both methods discussed above, the influence of the ground parameters and their spatial and temporal variations, as well as frequency dependence at the input impedances of the SRA and a calibrated antenna, were of primary importance. The other problem was created by the SRA load impedance and its temporal stability in the case of a thermistor and matching and attenuation in the case of microvoltmeter use. When an electromotive force at the input of a SRA is measured, all these factors are of secondary importance and they can be neglected in the accuracy estimations. In order to assure full precision of considerations as well as to present an opportunity to omit these factors, they will be considered. In the role of the emf detector, it is usual for diode detectors to be applied. By definition, the emf may be measured only with a device of infinitely high input impedance. In our case, the detector is represented by its input resistance R_d. Thus, the E-field intensity in the case discussed is given by:

$$E = V_A \frac{R_A + R_d}{R_d \, h_{eff}} \tag{4.23}$$

where: V_A – voltage measured at the SRA input.

A mean square error of the E-field assignment with the use of (4.23) (δ_E) is similar to that given by (4.21), but the role of the separate factors is different:

$$\delta_E = \sqrt{\delta_1^2 + \delta_2^2 + \left(\frac{R_A}{R_A + R_d}\right)^2 (\delta_3^2 + \delta_4^2)} \tag{4.24}$$

where: δ_1 – voltage measurement error,
   δ_2 – h_{eff} assignment error,
   δ_3 – SRA input impedance assignment or measurement error,
   δ_4 – detector input resistance measurement error.

The δ_1 error is estimated at the level of $\pm 3\%$ at frequencies below 500 MHz and not exceeding 4% at frequencies up to 1000 MHz. Here, HF voltage measurement is one of the most important factors limiting accuracy of the standard. A diode detector is usually calibrated at DC, and a frequency dependency may be observed.

Errors δ_2 and δ_3 have been discussed in section 4.4.1, and in this case it may be assumed that they are of a similar order; however, the influence of changes in the antenna input impedance is much less significant.

Error δ_4 is estimated at the level of approximately $\pm 2\%$. If it is assumed that the input resistance of a thin, half wave dipole, $R_A \approx 73\ \Omega$, and the input resistance of the detector, $R_d > 10$ kΩ, errors δ_3 and δ_4 can be neglected as their role does not exceed $\pm 0.01\%$. In practice, HF voltmeters of much higher input resistance are in use. However, any detector type represents an input capacitance and reactance X_d; the presence of a reactance may be neglected only in the case where $X_d \gg R_d$; if this condition is not fulfilled, the capacitance should be considered as part of a voltage divider, and the detector will measure a voltage other than the actual emf instead. The phenomenon may cause a small accuracy decrease in this type of standard when compared to that of the thermocouple detector.

The total inaccuracy of the EMF standard with diode detection is estimated at no more than approximately $\pm 5\%$ at frequencies below 300 MHz and $\pm 7\%$ up to 1000 MHz.

As in the case of the thermocouple detector, the diode detector is sensitive to temperature changes. Moreover, the changes are different at different points of the diode characteristics and, as a result, the role of temperature changes is different for different field intensities necessary in calibrations.

In summary, a diode detector is troublesome to use, but is actually the most favorable solution, and it is usually applied when primary SRA-type standards are of concern.

One comment more: As in the case of SRA-type standards presented in this chapter, and in any other cases presented in the book, any estimations of the standards' accuracy are done in relation to standards designed and constructed by the authors. However, there are no two identical designs, identical constructions, identical auxiliary equipment, nor identical environment; thus, every particular solution requires an individual approach to accuracy estimation. However, the discussions presented by the authors provide a good starting point to develop a meaningful understanding of the sources, natures, and likely magnitudes of associated errors.

4.5 Analysis of the accuracy of the STA method

As previously mentioned, the standard transmitting antenna method is based upon EMF generation using a transmitting antenna of known parameters (the STA), and the field intensity at a distance is calculated on the grounds of the parameters of the antenna, its excitation measurement, and the geometry of propagation between the

STA and an antenna of calibrated device or other OUT. The EMF at the location of the calibrated antenna is a result of interference between two rays (Figure 2.1), i.e., a direct ray and a reflected one. If the separation between the antennas is not large and the antennas are at low elevation, a surface wave should be considered as well. Under these conditions the E-field at the calibrated antenna is expressed by:

$$E = \frac{\varepsilon_0 \, f \, h_{eff}}{2} \sqrt{\frac{P}{Z}} \left[\frac{\exp(-jkr_1)}{r_1} + \frac{\exp(-jkr_2)}{r_2} \left[\Gamma + (1 - \Gamma) \, A \right] \right] \qquad (4.25)$$

where: P – power fed to the STA,
　　　Γ – reflection factor of the ground,
　　　A – surface wave attenuation factor.

This formula represents the situation where calibrations are performed on an OATS or when the reflected ray exists. When calibrations are done within an anechoic chamber, only the part representing the direct ray is different from zero in (4.25). Moreover, when the distance between the antennas exceeds 2λ, the presence of the surface wave can be neglected, i.e., $A = 0$. Thick half wave dipoles are often used as standard transmitting antennas, and the effective length of such a dipole is given by:

$$h_{eff} = \frac{\lambda}{\pi \, (1 - p)} \qquad (4.26)$$

where: p – h_{eff} increase factor, as shown in Figure 4.11.

Taking into account the above considerations and comments, the E-field is given by:

$$E = \frac{120}{1 - p} \sqrt{\frac{P}{Z}} \sqrt{\left(\frac{1}{r_1} - \frac{\Gamma}{r_2} \right)^2 + \frac{4\Gamma}{r_1 r_2} \sin^2 \left[2 \frac{k \, (r_1 - r_2)}{2} \right]} \qquad (4.27)$$

If the calibrations are performed on a terrain covered by a conducting mesh, or if it is possible to assume that the reflection factor $\Gamma = -1$ (this condition is usually fulfilled at frequencies above 30 MHz), and if the following condition is fulfilled:

$$\frac{h_t \, h_r}{d} < 0.5$$

then (4.27) may be simplified to the form:

$$E = \frac{120}{(1 - p) \, d} \sqrt{\frac{P}{Z}} \sin \left| \frac{2\pi \, h_t \, h_r}{d} \right| \qquad (4.28)$$

If, before calibrations start, the assumption that the reflection factor equals $\Gamma = -1$ should be checked, for instance, with the use of the method presented in section 4.3. If electrical parameters of the ground (σ and ε) are known, the reflection factor and its role in planned calibrations may be estimated using graphs, as shown in Figure 4.15.

Excitation measurement of the STA is the most troublesome, and it is a main source of calibration error. In older approaches a thermocouple at the antenna input was usually applied for the current measurement. A more profitable approach, currently preferred, is a measurement of the power fed to the antenna, as shown in Figure 4.16. The formulas above are based upon the latter solution.

A mean square error of the E-field calculation, for the H-field standard, with the use of the STA method, on the ground of (4.27), leads to the formula:

$$\delta E = \sqrt{\delta_1^2 + \delta_2^2 + \left(1 - \frac{\Delta\Gamma}{\Delta r}\right)^2 \frac{\delta_3^2}{4} + \left[1 - \frac{\Delta\Gamma + (r_1^{-1} + r_2^{-1})\Gamma k \sin k(r_2 - r_1)}{\Delta r}\right]^2 \delta_4^2}$$

$$(4.29)$$

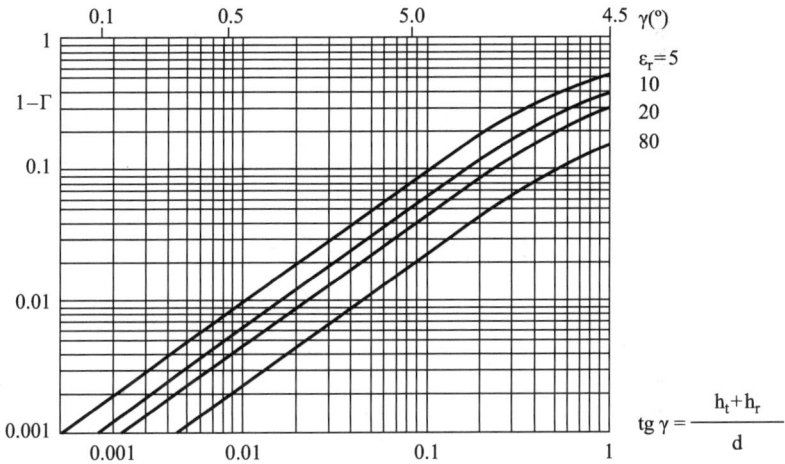

Figure 4.15   Reflection factor as a function of elevation angle γ for σ/ωε ≪ 1

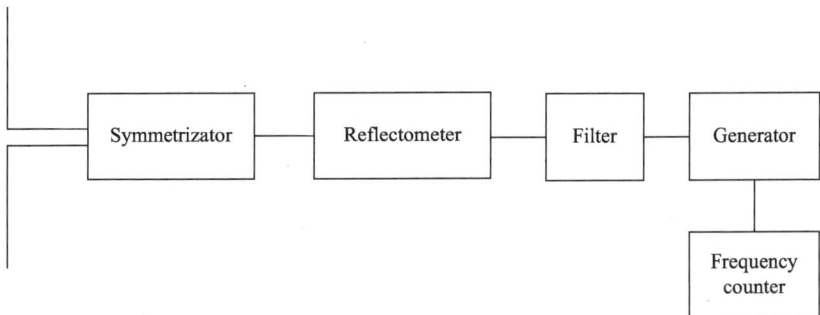

Figure 4.16   Excitation measurement of the STA

where: δ_1 – the STA excitation measurement error,

$\quad\quad\quad\delta_2$ – h_{eff} estimation error,

$\quad\quad\quad\delta_3$ – reflection factor measurement (estimation) error,

$\quad\quad\quad\delta_4$ – an error of linear sizes measurements, i.e.:

$$\delta_4 = \frac{\Delta r_1}{r_1} = \frac{\Delta r_2}{r_2} = \frac{\Delta l}{l} \tag{4.30}$$

and auxiliary terms in (4.29) are:

$$\Delta\Gamma = r_1^{-2} + \Gamma^2 r_2^{-2} \tag{4.31a}$$

and

$$\Delta r = (r_1^{-1} + \Gamma r_2^{-1})^2 + \frac{4\Gamma}{r_1 r_2} \sin^2 \frac{k\,(r_1 - r_2)}{2} \tag{4.31b}$$

An error in the wavelength (frequency) measurement was neglected in the above considerations because it is much less than any other factor limiting the accuracy of the calibration.

It may be noticed that the value given by (4.29) does not exceed the following:

$$\delta_E \leq \sqrt{\delta_1^2 + \delta_2^2 + \delta_3^2/4 + \delta_4^2} \tag{4.32}$$

The error δ_1 includes a class of the applied reflectometer (measurement error declared by its manufacturer at given frequency) and inaccuracy of estimations of losses introduced between the reflectometer and the STA by symmetrization and matching the network. In order to neglect the influence of the ground and mismatching of the STA due to it, the antenna should be elevated as derived from the curves in Figure 4.13. The excitation measurement in the case of the STA method is less troublesome and more accurate compared with the voltage (emf) measurement at the input of an SRA, which results mainly from the much higher levels of the measured signals and the possibility of applying typical equipment. It is estimated that the error should not exceed $\pm 1\%$ at frequencies up to 1000 MHz.

Error δ_2 has already been discussed, and its value is similar to that of the SRA, i.e., $\delta_2 \leq \pm 2\%$. This illustrates, by the way, the capability of using the same antenna as an SRA and an STA.

Error δ_3 may dominate the accuracy of the STA standard. In order to limit the influence of the ground parameters and ensure $\Gamma = -1$, the ground should be covered by a conducting mesh. In the case of calibrations performed at an OATS, and the necessity of measuring the ground parameters, the value of the error may reach $\pm 5\%$.

Error δ_4 represents distance measurement accuracy (4.30). It may be assumed that it does not exceed $\pm 0.5\%$. However, when calibrations are performed at an OATS without use of any means to improve its flatness, the error may even reach $\pm 1\%$.

The above considerations do not take into account an ambient temperature influence upon STA excitation measurements. The factor here is similar to that in the case of SRA. It is a systematic error, and its value is usually known *a priori*, which makes it possible to take it into account when results of measurements are interpreted.

In the above considerations, the error caused by nonparallel placement of the antennas and displacement of their symmetry axes was not taken into account. This results in generation of the standard EMF with estimated accuracy, however, not in the place of the calibrated antenna. The additional error due to this phenomenon should not exceed 1%, and it can easily be detected when calibrations are repeated. The first proof of correct antenna placement is maximal indication of the calibrated device. If the maximal indication does not coincide with parallel placement of the antennas, this may indicate that side reflections not taken into consideration may be occurring.

The maximal error of the calibration method using the STA method should not exceed ±4–5% in the frequency range 30–1000 MHz.

The above considerations show that the STA method ensures accuracy on a level similar to the SRA method. With all the methods discussed, is possible to improve their accuracy, especially as a result of more accurately taking into account the electric parameters of the ground and calculation (measurement) of the reflection factor.

An evident increase accuracy is obtainable by way of total elimination of the ground parameters' influence, i.e., a placement of the standard in an anechoic chamber. As has already been mentioned, this solution is irreplaceable in every respect, with one exception: the cost of the solution, especially at the lowest frequencies or when large object are to be tested. The worst-case performance is discussed here, which may be useful when accuracy estimation are performed for any standard.

To illustrate differences in accuracy estimated for standards placed at an OATS with those placed within an anechoic chamber, below are the set-up errors of the above-discussed standards when applied within a chamber:

- SRA standards: with thermocouple detector: $\delta_E \leq \pm 4\%$,
  with diode detector: $\delta_E \leq \pm 4.5\%$,
  with selective microvoltmeter: $\delta_E \leq \pm 6\%$,
- STA standard: $\delta_E \leq \pm 3.5\%$.

Again, these are based on the authors' experiences but serve to highlight the errors that are likely in well-designed and -executed measurements.

4.6   Directional antenna calibration

The expression "directional antenna" may suggest that there exists an antenna that is "omnidirectional." Such a device is still unknown, with the exception of omni-directional EMF probes, used for EMF quantification for environmental purposes.

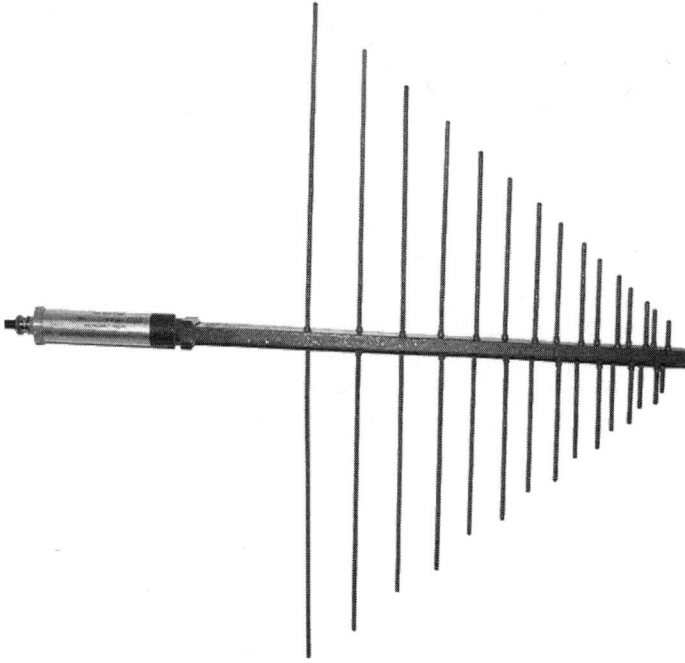

Figure 4.17   DAMZ-4/50 log-periodic antenna

Applied in EMF metrology, dipole antennas, loop and ferrite ones, are described by their radiation pattern. Electrically small antennas are of sinusoidal radiation pattern (without regard to their number in a system and manner of their connections), while resonant dipoles have patterns of similar shape, with some gain in the relation to the former. We may generalize that their directivity is relatively small, and in a cross section, the pattern is circular. A directional antenna is usually understood to be an antenna containing several individual elements. The combination of these elements may be arbitrary. It may contain a number of active elements (e.g., phased arrays, log-periodic antennas), or active and passive elements (e.g., Yagi-Uda antennas). One class of directional antennas includes spiral antennas and horn ones. The gain of the antennas is a function of their electrical sizes, while their wideband properties are specific to different types.

A good combination of directional properties and wideband nature is represented by a log-periodic antenna (LP). As a result, this is one of the most popular antennas used in EMF metrology within the frequency range (10) 300–1000 (2000) MHz.

In the following discussions, we take as an example a DAMZ-4/50 LP antenna manufactured by INCO in Poland (Figure 4.16) [43]. This antenna is designed to work within the frequency range 300–1000 MHz. Its gain, in relation to a half wave dipole, is 7 dB at 500 MHz and its front to back ratio is −20 dB. A directional diagram of the antenna in the E- and H-planes is shown in Figure 4.18. For the purpose of further analysis, the directional diagram of the antenna in the H-plane,

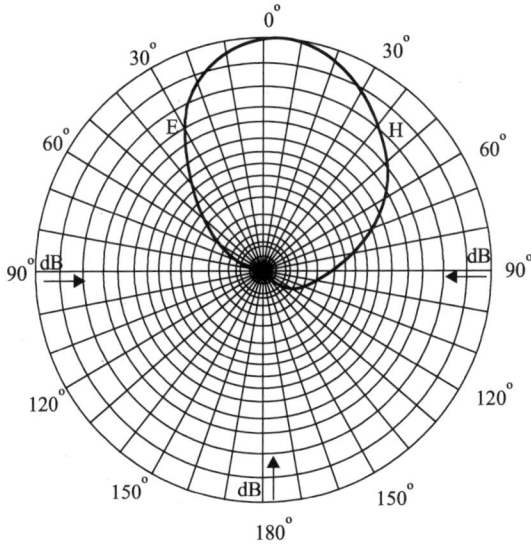

Figure 4.18   Directional pattern of the DAMZ antenna in E- and H-planes

with an accuracy better than ± 1 dB, for elevation angles $\varphi < 70°$, for main lobe direction $(\Theta = 0)$, is approximated by the equation:

$$f(\Theta, \varphi) = \frac{1 + \cos^2 \varphi}{2} \qquad (4.33)$$

Of course, the pattern of a log-periodic antenna is a function of its length, number of elements, obtuse angle, frequency, etc. However, the primary role at any frequency is played by three adjacent dipoles, and the pattern of the antenna is similar to the three-element Yagi-Uda. Assumed above for consideration, a type of antenna and its pattern approximated by (4.33) introduces some limitations regarding their generality, but it does not limit the generality of the conclusion that can be drawn on the ground of these considerations.

The methodology of calibration is similar to that applied to the dipole antennas, and here, both the methods can be used, i.e., STA and SRA. We will assume the SRA method for the present; these considerations may be repeated for the STA method, and they lead to similar conclusions. An E-field in the place where an SRA is, and then replaced by an antenna under test, is given by (4.25). If the presence of the surface wave can be neglected, and assuming that the condition $d > 2D^2/\lambda$ is fulfilled, then the formula may be rewritten in the form:

$$E = A \left(\frac{\exp(-jkr_1)}{r_1} + \Gamma \frac{\exp(-jkr_2)}{r_2} \right) \qquad (4.34)$$

where: A – amplitude, which is not essential here.

When using the SRA method for calibration, and the calibration is performed at frequencies above 300 MHz, it is profitable to use, as a transmitting antenna, a directional one as well. In order to simplify our considerations, we will assume that the role may be played by identical antennas as calibrated and that the two antennas are placed at the same height. In this case, the E-field given by (4.34) is modified to give:

$$E = A \left(\frac{\exp(-jkr_1)}{r_1} + \Gamma \frac{\exp(-jkr_2)}{r_2} f(\Theta, \varphi) \right) \tag{4.35}$$

The error of calibration of a directional antenna contains two components: one caused by E-field inhomogeneity along the antenna (due to its second dimension, length) and the other by directional properties of the antenna. We will consider them both.

4.6.1   E-field averaging along the antenna

In order to show the role of the effect of E-field averaging along the antenna, we will assume that the calibration is performed in the presence of only a direct ray. Of course, all previously mentioned conditions typical of the SRA method should be preserved. A relative change of E-field at the length of the antenna (D) is:

$$\left| \frac{\Delta E}{E} \right| = \frac{D}{d} \tag{4.36}$$

Total length of the DAMZ antenna is 70 cm; thus, in order to have E-field homogeneity at the length of the antenna better than 1%, it would be necessary to perform calibrations at distance $d = 70$ m. If we assume an acceptable inhomogeneity of the field at the level of $\pm 1\%$ in relation to the antenna center, the distance decreases to half. As mentioned above, at any frequency only three adjacent dipoles of the LP antenna are active. If we assume that the length of the active part of the antenna does not exceed 0.25λ and the requirement related to the field homogeneity should be limited to this part of the antenna, we will have, for the longest wave at which the antenna works, an active length of around 25 cm; then the requirement of the minimal distance d decreases to about 2 m (that is exceptional, but may be used). In the first approximation, at the length of the calibrated LP antenna, the E-field is a linear function of d. Thus, at the active length of the antenna, the averaged E-field is an arithmetic mean of the intensities at the beginning and at the end of the length and is equal to E-field calculated for its center. If we assume that during calibration a spatial location of the SRA should be precisely the same as the center of the active part of the calibrated LP antenna, then the role of the E-field inhomogeneity along the antenna may be neglected. This conclusion was experimentally proven during field measurements performed at distance 1 m between a half wave dipole transmitting antenna and the LP antenna considered, even at the lowest frequencies.

4.6.2 Role of the radiation pattern

Here are three combinations of the transmitting and receiving antennas:

- The simplest case, equivalent to the calibration method of a dipole antenna, i.e., both antennas, the transmitting and the receiving one, are half wave dipoles; in this case the reflected component at the point of observation E_r is identical to that at the right side of the brackets in (4.34):

$$E = A \frac{\exp(-jkr_2)}{r_2} \Gamma \tag{4.37}$$

- One of the antennas (arbitrary choice; the approach may be applied when dipole antennas are calibrated, in order to increase the E-field intensity) is a half wave dipole and the other a directional one; then the reflected component of the E-field at the point of observation is equal to the right part of the brackets in (4.35), i.e.:

$$E = A \frac{\exp(-jkr_2)}{r_2} \Gamma f(\Theta, \varphi) \tag{4.38}$$

- Both antennas are directional; then the reflected part of the resultant E-field at the point of observation is:

$$E = A \frac{\exp(-jkr_2)}{r_2} \Gamma f^2(\Theta, \varphi) \tag{4.39}$$

The E-field in a point of observation is a superposition of the direct ray (E_d) and the reflected one (E_r):

$$|E| = \sqrt{|E_d|^2 + |E_r|^2 + |E_d||E_r|\cos(\Phi_r - \Phi_d)} \tag{4.40}$$

where: Φ_d and Φ_r – the phase of the direct and the reflected rays, respectively.

Now we assume, for simplification only, that both antennas are at the same height $h_t = h_r$, and thus $r_1 = d$; and now we calculate a ratio of the resultant E-field, E, to the direct component in the case when one of the antennas is a dipole and the other is a LP one:

$$\frac{|E|}{|E_d|} = \sqrt{1 + \left(\frac{d}{r_2}\right)^2 \left(\frac{1 + \cos^2\varphi}{2}\right)^2 |\Gamma|^2 + \frac{d}{r_2}|\Gamma|(1 + \cos^2\varphi)\cos\Delta\Phi} \tag{4.41}$$

where: $\Delta\Phi = \Phi_r - \Phi_d$ – phase difference between direct and reflected ray.

This equation makes it possible to accurately calculate the field distribution in the direction of the transmitting antenna's radiation maximum. However, the calculations would be of limited use for further considerations. Assume now that the reflection factor $\Gamma = -1$ and, in order to find maximal differences, $\Delta\Phi = n\pi/2$

(where: n – integer). Now we can calculate envelopes of all curves that could be obtained from (4.41), and similar ones for all three cases considered.

An envelope O_{dd}, while the both antennas are half-wave dipoles is:

$$O_{dd} = \frac{|E|}{|E_d|} = 1 \pm \frac{d}{r_2} |\Gamma| \qquad (4.42)$$

For the case of a half wave dipole and an LP antenna, the envelope O_{dl} is:

$$O_{dl} = \frac{|E|}{|E_d|} = 1 \pm \frac{d}{r_2} \frac{1 + \cos^2\phi}{2} |\Gamma| \qquad (4.43)$$

and for the case of two LP antennas, the envelope O_{ll} is:

$$O_{ll} = \frac{|E|}{|E_d|} = 1 \pm \frac{d}{r_2} \left(\frac{1 + \cos^2\phi}{2}\right)^2 |\Gamma| \qquad (4.44)$$

The calculated values of the envelopes for $h_t = h_r = 2$ m and $\Gamma = -1$ as a function of d are shown in Figure 4.19.

The aim of these considerations is to find the accuracy of the directional antenna calibrations using the SRA method. The influence of the directivity results from the presence of the reflected ray. The resultant field at a point of observation seen by the SRA (always half wave dipole) and a calibrated antenna (in our case an LP) is different and leads to an additional calibration error. In order to minimize the error, we will assume that the calibration is performed in the maxima of the field, i.e., we will take into account the upper curves in Figure 4.18.

The maximal value of the additional calibration error is the difference between two curves, O_{dd} and O_{dl}, when a dipole and LP are in use, and the error η_d is:

$$\eta_d = \frac{O_{dd} - O_{dl}}{O_{dd}} = \frac{d}{2} \frac{\sin^2\varphi}{d + r_2} \qquad (4.45)$$

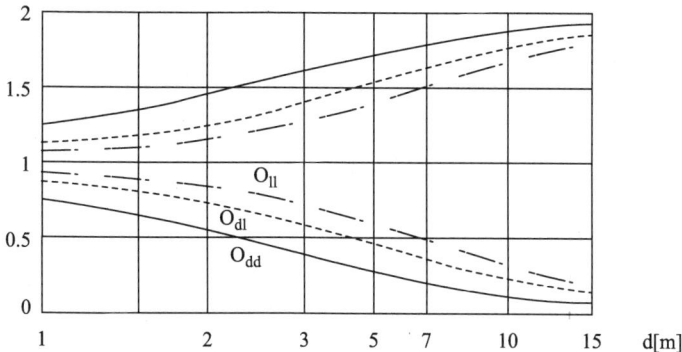

Figure 4.19   Envelopes O_{ll}, O_{dl}, and O_{dd} as a function of d

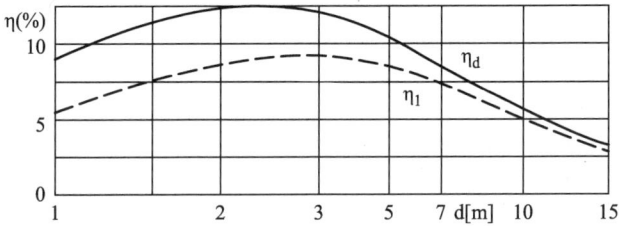

Figure 4.20 The errors η_l and η_d as a function of d

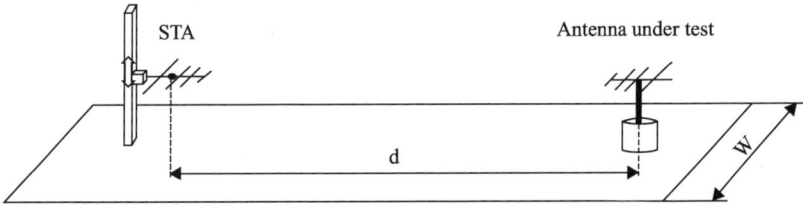

Figure 4.21 A set-up for automated directional antenna calibration

and, similarly, η_l when the transmitting and the calibrated antennas are identical log-periodic antennas:

$$\eta_l = \frac{O_{dl} - O_{ll}}{O_{dl}} \qquad (4.46)$$

Calculated values of the two errors are shown in Figure 4.20. The errors represent maximal values of additional error that appears while the SRA method is applied for an LP antenna calibration on an OATS. The error should be taken into account when total inaccuracy of the calibration is estimated. These considerations may be summarized as follows:

- It is fully possible to use the calibration methods designed for dipole antennas for the calibration of directional ones based on testing at an OATS.
- Additional errors of such a procedure are relatively small, especially when larger distances d are allowed (acceptable when high-sensitivity EMF meters are calibrated); moreover, there is a possibility to improve calibration accuracy by using high-gain antennas as transmitting antennas as well as using calculated correction factors.
- The calculations presented here are an example, and nothing more. They were performed for assumed types of antennas and calibration conditions. However, they show how the calculations may be performed when other antennas and conditions are of concern.
- All these procedures are unnecessary when a calibration is performed in an anechoic chamber; in this case, the additional errors vanish.
- Although it is possible to correctly calibrate an EMF meter with a directional antenna, this does not mean that it ensures the required accuracy in any case (see Figure 2.3).

An example of a set-up for automatic log-periodic antenna calibration at the OATS, applied by the authors, is shown in Figure 4.21. The configuration contains a log-periodic standard transmitting antenna and, at distance $d = 10$ m, an antenna under test. The standard log-periodic antenna was previously calibrated using a standard receiving antenna, which makes it possible to use the system for calibration of any type of dipole and directional antennas. This set-up was used within a frequency range of 100–1000 MHz, allowing measurements of the antenna factors and radiation patterns of any antennas and ensuring full automatization of the measurements. However, in reference to the above-discussed procedures, this approach, although very useful and making measurements quite quickly, accurately, and easily, represents a secondary standard rather than a primary one.

Chapter 5

Accuracy analysis of EMF standards
with loop antennas

At frequencies below 30 MHz E-field meters with loop antennas are in general use for measurements in the far field. Although the meters measure the H-field, they are calibrated in E-field units (μV/m, dBμV/m). But their calibration must be done with a standard magnetic field source; this leads to the need to use (3.3) during the calibrations.

A similar exception in the loop antenna standard application for E-field meter calibration was discussed in section 3.2.3. Here, only the accuracy of H-field generation will be considered. In these considerations, only "internal" factors will be taken into account, i.e., location of a transmitting and a receiving antenna, excitation measurement error, electrical sizes of the antennas, and others. Because of the specificity of the standard, it is possible to neglect here any "external" factors limiting accuracy of the standard, such as the influence of ground conductivity, the presence of other conducting objects in the neighborhood of the standard, multipath propagation, or interference caused by external EMF. This simplification is acceptable because of the fact that the antennas are much smaller compared to the wavelength [47] and, as a result, are much less sensitive to the factors mentioned. For the same reason, the H-field standards may work in a less rigorously controlled environment, with no screened or anechoic chambers.

In general, H-field metrology is performed at frequencies below 30 MHz in far-field metrology, and well above 30 MHz for near-field measurements for labor safety and environmental protection purposes. The difference is mainly in measured field levels. However, as has already been mentioned, especially in relation to the near field, wideband H-field sensors, probes, and meters, an ability to check their frequency response over a much wider frequency range is required.

5.1 Accuracy analysis of the SRA standard

A calibration procedure for SRA method applications for loop antennas is presented in section 3.2.2. A block diagram of a SRA applied by the authors is shown in Figure 5.1. This setup contains a loop antenna loaded with a diode detector, and the resulting DC is fed to a DC millivoltmeter through a low-pass filter.

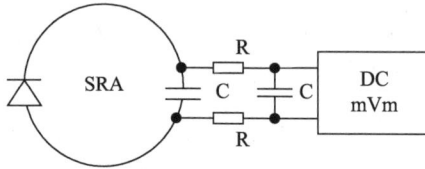

Figure 5.1   Block diagram of a standard receiving loop antenna

The accuracy of the standard analysis is based upon (3.5). The formula contains only three variables that can limit accuracy: e_A, f, and S_r. Their role is briefly analyzed below.

5.1.1   Error in e_A measurement

Because of relatively low input resistance (R_A) in the electrically small loop, it is possible to use voltmeters of much lower input impedance compared to those used in the case of dipole antenna calibration. This means that the set is less sensitive to external EMI, and this factor can be neglected in the analysis.

Because of the resistance and the much lower frequency range compared with the dipole standards, it is possible to assume that the error in e_A measurement (δ_1) does not exceed ±2% at frequencies below 30 MHz. However, it increases substantially at higher frequencies.

There is another possibility, namely to measure the current, I, in the antenna instead of the emf. In the majority of cases the following condition is fulfilled: $\omega L_A \gg R_A$ (where: L_A – inductance of the antenna); thus, (3.5) may be transformed as:

$$H = \frac{I L_A}{\mu_0 S_r} \tag{5.1}$$

where: μ_0 – permeability of the vacuum (free space).

This equation shows the possibility of using current measurement in the antenna instead of EMF.

There are two possibilities:

- Use of a sensitive thermocouple is sometimes applied in H-field sensors; however, this solution places a major limitation on the dynamic range of the measurements.
- Using a current transductor to measure the current, contrary to other solutions, directly at high frequency with a use of a selective voltmeter.

5.1.2   Error in loop surface area measurement

It is possible to assume that linear sizes are measured with an inaccuracy below 1%; thus, the surface area of the antenna should be measured with an error δ_2 not exceeding 2% at frequencies below 100 MHz. The error increases at higher frequencies as a result of the necessary limitation of the size of the antenna used. In the error estimation, the finite diameter of the antenna conductor, electrostatic screen (if applied), and other factors are taken into account. All of them require individual approaches to their estimation.

5.1.3 Error in frequency measurements

Error in frequency measurement (δ_3) is much lower compared to the role played by other factors. Even if the frequency is read from an analog scale of an older type generator, the error is unlikely to exceed 0.01%. Modern generators ensure frequency readings with an accuracy to several significant figures. This factor is mentioned only for formal reasons as frequency appears in (3.5), among others.

5.1.4 Other factors limiting the standard's accuracy

The above-discussed factors are represented in (3.5); however, there are several other factors that could affect accuracy, although they are not directly represented in the formula.

5.1.4.1 Error due to nonparallel H and S_r vectors

A scalar product of the vectors H and S_r is a function of the cosines of the angle between them. If it can be assumed that the inaccuracy of the antennas' positioning may reach 5°, then, if we assume square law characteristics of the detection diode, the error δ_4 representing noncoaxiality of the vectors may be assumed to not exceed $\pm 1\%$. Thus, the mean square value of the calibration error δ_H is:

$$\delta_H = \sqrt{\delta_1^2 + \delta_2^2 + \delta_3^2 + \delta_4^2} \tag{5.2}$$

If we take into account the estimated values of the separate errors, the error of calibration will be $\delta_H \leq \pm 3.5\%$ at frequencies below 100 MHz and $\delta_H \leq \pm 4.5\%$ at frequencies up to 200 MHz. It is necessary to remember that (5.1) was introduced with some simplifications that decrease the estimated accuracy. Taking these into account and rounding, we may suppose that the inaccuracy of the discussed standard should not exceed $\pm 4\%$ below 100 MHz and $\pm 5\%$ up to 200 MHz, and that the inaccuracy will increase further with increasing frequency.

5.1.4.2 Error due to the resonant effects in the SRA

If, at frequency f_0, there appears a resonance of the SRA inductance and capacitances of the antenna and other (dispersed) capacitances of the set, for frequencies $f < 0.5 f_0$ we may write:

$$\left| \frac{e'_A}{e_A} \right| = \left[1 - \left(\frac{f}{f_0} \right)^2 \right]^{-1} \tag{5.3}$$

where: e'_A – emf induced in SRA and increased due to resonant phenomena.

It may be estimated that if the error due to the resonant emf increases, it should not exceed 1%; the SRA should not be used at frequencies $f > 0.1 f_0$.

If this condition is not fulfilled, it is necessary to undertake additional calibration of the SRA, recording appropriate correction factors that would allow the antenna to be used at higher frequencies. The possibility exists here because the error is of a systematic character. Although the use of the correction factors may be troublesome, this approach makes it possible to widen the frequency range at which the SRA may be applied. At low frequencies the problem almost does not exist (with the exception of multi-turn loops); however, it is of primary importance at higher frequencies due to technological problems with constructing smaller and smaller antennas.

5.1.4.3   Error due to the "antenna effect" in the SRA

In antenna theory, the "antenna effect" describes the phenomenon of a nonscreened loop antenna being sensitive to the electric field as well. The effect is applied, for instance, in the construction of power density probes, where multiple loaded loop antennas are used. In the case of standard antennas, the effect is undesirable and may lead to large errors. At lower frequencies, it is possible to protect a loop against the effect by electrostatic screening of the antenna. However, at higher frequencies, where the sizes of the loop have to be maximally reduced, for example, due to the resonant phenomena, screening may create unsolvable problems, and then nonscreened loops are used. The relation of emf e_E induced in a circular standard loop of radius r_r by an electric field in relation to emf e_A induced by a magnetic field is given by:

$$\left| \frac{e_E}{e_A} \right| = \frac{4\pi\, r_r}{\lambda} \qquad (5.4)$$

This equation was introduced for plane wave conditions where, in order to eliminate the fulfilling condition $e_E < 0.01e_A$, the following condition should be met: $2r_r < 1.6\ 10^{-3}\lambda$. This condition is fully acceptable for frequencies below 30 MHz. However, even at these frequencies, it may lead to a reduction of SRA size and, as a result, sensitivity. At frequencies above 100 MHz, the condition is almost unrealiazable. Moreover, the condition is valid for the plane wave where the relation E/H is known and constant. In the conditions of the near field, the relation E/H may be arbitrary. Close to a loop transmitting antenna, there appears a strong E-field, as was presented in section 3.2.1. If we have some *a priori* data regarding the nature of the field in which calibrations are to be performed, it is possible to estimate the role of the phenomenon using:

$$\left| \frac{e_E}{e_A} \right| = \frac{4\pi r_r}{\lambda} \left| \frac{Z_t}{Z} \right| \qquad (5.5)$$

where: Z_t is an impedance of the field correlated with a given EMF source and defined as the ratio E/H at a point considered.

For two identical and coaxially placed circular loop antennas, if $2r_t < 0.2d$, the formula may be rewritten in the form:

$$\left| \frac{e_E}{e_A} \right| = \left(\frac{4\pi\, r_r}{\lambda_0} \right)^2 \qquad (5.6)$$

In practical applications, if the antennas are small in relation to the shortest wavelength at which they work, and not tuned to resonance by added reactances, in order to have the fulfilled condition, $e_E < 0.01e_A$ is enough if $2r_t < 1.6.10^{-2}\lambda$.

The above discussion shows that the "antenna effect" may be neglected in the majority of cases, especially when calibrations are performed at low frequencies. However, it shows a necessary caution when higher frequencies are in use or if the relation E/H of a source is unknown. In any case, there is a possibility for experimental verification of the presence of the error and its value by turning an SRA, antenna under test, H-field sensor, or other device through 180°. An error δ_5, representing the phenomenon, we will define in the form:

$$\delta_5 = \frac{\alpha_1 - \alpha_2}{\alpha_1 + \alpha_2} \tag{5.7}$$

where: α_1 and α_2 – indications of a calibrated device before and after turning it through 180°.

5.1.4.4 Error resulting from radii differences between the SRA and an antenna under test

An error in calibration using the SRA method and resulting from different radii between the SRA and an antenna under test was discussed in section 2.5. We only recall it here in case the difference may play an important role in the estimation of the accuracy of a SRA standard. This source of error is especially important when a transmitting antenna, an SRA, and an antenna under test are placed in close proximity, perhaps to maximize the H-field intensity.

5.2 Accuracy analysis of an STA standard

This method has been presented in section 3.2.1, while the field inhomogeneity around a transmitting loop antenna has been presented in section 2.5. In this discussion, we will refer to (2.37), which is most often applied in the calibrations. As may be deduced from the equation, the H-field intensity accuracy, in the case discussed, is limited by the accuracy of the antenna excitation measurement and the accuracy of the measurement of the geometric configuration (size of the STA, size of an antenna under test, distance). The values of separate errors are estimated below.

The block diagram of a version of the H-field standard designed by the authors is shown in Figure 5.2. This set-up includes an STA, a thermocouple for current measurement, a power source, and a DC milivoltmeter.

5.2.1 Accuracy of the current measurement

The most frequently used device for current measurement is a thermocouple. The accuracy of the thermocouple calibration at DC and its use in HF current measurement was considered in section 4.4.1; we may assume here that the error of the current measurement is of similar order, i.e., $\delta_1 \leq \pm 2\%$.

5.2.2 Accuracy of linear size measurement

This factor was discussed in section 4.5, and the estimation of the error given by (4.30) is still valid. However, in the case of the H-field, it is represented in two

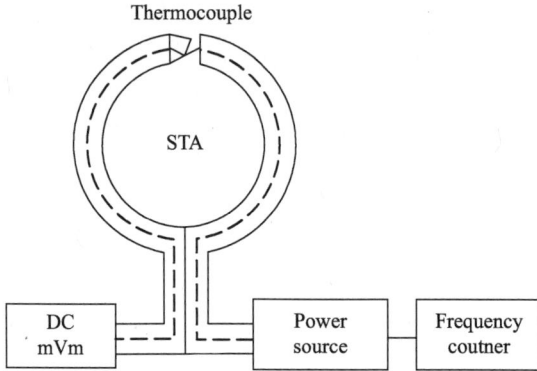

Figure 5.2   STA H-field standard used by the authors

ways: as a measure of r_t measurement accuracy (and then calculation of S_t), as well as d, r_r, and r_c, necessary to calculate D [(3.7) and (3.8)]. It is possible to assume that the error of the linear size measurement in the case discussed is $\delta_2 \leq 3\Delta l/l$ (where: Δl – error of the linear size measurement). It is possible to assume that the error does not exceed 1% at frequencies below 100 MHz and $\pm 1.5\%$ up to 200 MHz. At higher frequencies, the error increases further due to the necessity of constructing smaller and smaller antennas that may create technological problems in their accurate manufacture and dimension measurement (taking into account diameter of the conductor the antenna is made of).

5.2.3   Error due to noncentric placement of the antennas

The value of the error due to noncentric placement of antennas δ_3 is estimated to not exceed $\pm 1\%$. Thus, the mean square value of the calibration error, estimated on the ground of (2.37), is given by:

$$\delta_H = \sqrt{\delta_1^2 + \delta_2^2 + \delta_3^2} \qquad (5.8)$$

If it is assumed that the levels of separate errors are as discussed above, we may estimate a mean square error of the calibration at the level of $\delta_H \leq \pm 3\%$ at frequencies below 100 MHz, and $\delta_H \leq \pm 4.5\%$ at frequencies not exceeding 200 MHz. Rounding these values, due to unexpected errors, we will have $\delta_H \leq \pm 4\%$ below 100 MHz and $\delta_H \leq 5\%$ up to 200 MHz.

The above discussion does not take into account that (2.37) was introduced with some simplifications, which leads to the accuracy of the H-field calculated on this basis having an error of 1%. Two comments on the simplifications:

- The simplifications are of deterministic character, and they could be taken into account when calibrations are performed, or, if better accuracy is required, more terms of (2.36) should be taken into consideration.

- If we take the error of (2.37) into account and we add it to (5.8), we will see that its role is negligible.

5.2.4 Other factors limiting the standard's accuracy

The above discussion took into account only magnitudes represented in (2.37). However, there are other factors limiting the accuracy of the standard that are not represented in the equation, as they were neglected during its introduction or they were not taken into account when introductory assumptions were made. We present a few of them below.

5.2.4.1 Nonuniform current distribution along the STA

When (2.30) was transferred to (2.32), it was assumed that the current along the STA was constant, which allowed the exclusion of the current, I, from the integration. There are two reasons that the condition $2r_t \ll \lambda$ may not be fulfilled:

- a necessity to generate H-field as strong as possible and resulting application of multi-turn loops,
- a necessity to work at highest possible frequencies.

In both cases the condition may not be fulfilled, in which case the current distribution along a loop is sinusoidal:

$$I = I_m \cos \gamma l \tag{5.9}$$

where:

　　l – distance to a current maximum,
　　I_m – current intensity in the maximum,
　　I – current at distance l from the maximum.

In order to perform a quantitative estimation of the error due to nonuniform current distribution along the antenna and its possible role in degradation of the accuracy of calibration, the error δ_{nu} will be defined in the form:

$$\delta_{nu} = \frac{|H_{nu}| - |H_{AV}|}{|H_{nu}|} \tag{5.10}$$

where:

　　H_{AV} – average H-field calculated along with (2.37),
　　H_{nu} – average H-field calculated taking into account nonuniform current distribution.

$$H_{nu} = \frac{\sqrt{1 + k^2 R_0^2}}{2\pi R_0^3} \int_l I \, dl \tag{5.11}$$

If (2.37) and (5.9) are substituted into (5.10) and then (5.11), we will have:

$$\delta_{nu} \approx \frac{\pi^2}{6}(k\,r_t)^2 \tag{5.12}$$

If we assume, for instance, that the error $\delta_{nu} \leq 0.6\%$, then we will have for a single turn, circular loop antenna the following condition:

$$r_t \leq 0.01\lambda \tag{5.13}$$

This formula presents the maximal radius of the STA, while an error due to non-uniform current distribution in it should not exceed 0.6%. This is only an example; the role of the error is much less if we take into account other sources of standard inaccuracy [see (5.8)]. However, it illustrates well the need for precise error estimations when the SRA is to work at relatively high frequencies or when an application of multi-turn loops is planned.

5.2.4.2   Finite sizes of a calibrated antenna

Equation (2.37) gives the averaged value of the magnetic field intensity at the surface of a calibrated antenna. It is easy to see that the value is identical to the modulus of the radial component of the H-field radiated by a small loop, as given by (2.28). Because of the finite size of the set, the calibrated antenna is partly sensitive to the transverse component of the transmitting antenna, given by (2.29). Thus, the electromotive force induced in the antenna is:

$$H = 2\int_0^\Theta (H_r + H_\Theta)\, d\Theta \tag{5.14}$$

where: Θ – an angle at which the calibrated antenna is seen from the center of the STA (see Figure 5.3):

$$\Theta = \text{arc tg}\, \frac{r_c}{d}$$

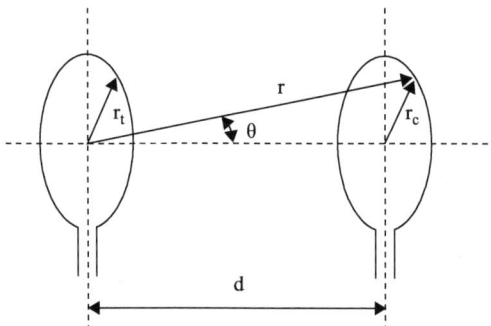

Figure 5.3   STA and the calibrated one

An error due to finite size of the calibrated antenna, δ_f, can be defined in the form:

$$\delta_f = \frac{|H| - |H_{AV}|}{|H|} \tag{5.15}$$

For small values of the angle Θ we have:

$$\delta_f \leq \frac{1}{6}\Theta^2 \tag{5.16}$$

The error δ_f appeared in (2.37) due to simplifications applied during its introduction, and it may be eliminated when more terms of (2.36) are in use. However, the estimations presented focus attention on the presence of the H-field components radiated by the STA and show the need to use larger distances d if at all possible. We may add here that the value of δ_f given by (5.16) is overestimated because, during its introduction, phase changes at the surface of the antenna were not taken into consideration.

5.2.4.3   Nonlinear distortions of STA excitation

At the very beginning of this discussion it was assumed that any field source is excited by a monochromatic signal. From a theoretical point of view, this was absolutely correct. However, in practice, every power source introduces nonlinear distortions and, as a result, its output signal is more or less a multichromatic one [19]. The effect is negligible when good quality signal generators are used, as for "low-level" meter calibration. In the case where strong fields have to be generated, linear amplifiers are usually used for signal amplification. Some nonlinear distortion is usually acceptable when working with output power not exceeding medium levels. When they are used at their maximal levels (and sometime above them), the distortions may reach substantial levels regardless of the type of amplifier, its manufacturer, etc. Because of this, the presence of distortions should be taken into account, especially when maximal field levels are necessary in calibrations or other field source applications. Although during calibrations, or other work of a technical nature, it is usually possible to check the spectrum generated by our field source, in the case of biomedical investigations such a possibility is not taken into account; as a result, in these experiments accords of sine waves are applied instead. Years ago it was discovered by Prof. Garkavi from Rostov [15] that the biological activity of accords due to synergetic effects is completely different as compared to the sine wave. This may be one of the reasons why bioelectromagnetic experiments performed in "identical" conditions in different centers give different results.

As previously mentioned, the effect of distortions appears in any type of standard STA as SRA ones. Considerations related to the role of harmonics are included in the discussion of H-field standards because they are the most sensitive to the presence of the effect. The sensitivity results from the frequency dependence of radiation efficiency and effective height of loop antennas. This increases with frequency, which leads to passive amplification (a "favorization") of higher harmonics of a standard exciting source.

Because of nonlinear distortions in power amplifiers and insufficient output signal filtration, or even a deficiency in such a procedure, the output current I of a generator, exciting an EMF standard, is a sum of fringes, especially at harmonic frequencies:

$$I = \sqrt{\sum_{n=1}^{\infty} I_n^2} \tag{5.17}$$

where: n – order of a harmonic.

As a result, the generated H-field contains n harmonics:

$$H = \sqrt{\sum_{n=1}^{\infty} H_n^2} \tag{5.18}$$

where: H_n – averaged at the surface of receiving antenna H-field at n-th harmonics:

$$|H_{AV}|_n = \frac{I_n S_t}{2\pi R_0^3} \sqrt{1 + n^2 k^2 R_0^2} \tag{5.19}$$

This formula takes into account a frequency dependence of the parameters of an STA. The subject of calibration can be a magnetic antenna of a selective meter, a wideband H-field probe, or a standard receiving antenna. The frequency response of these will be different, and thus their response to an achromatic field is different. Let's summarize: A selective meter is able to select a single frequency fringe from the whole spectrum, a wideband H-field probe has a frequency response that is flat over a frequency range, while h_{eff} of an SRA increases with frequency [see (3.5)]. Moreover, the response to the field is a function of detection type and is different for rms or peak value detectors.

Below, we consider six cases of errors caused by the effect discussed, when three types of meters with two types of detectors are applied. In order to make them clear, the errors have two-letter indexes. The first letter indicates the type of calibrated meter (s-selective, w-wideband, a-SRA), and the second indicates the type of detection (p-peak, e-rms).

a.   Selective meter calibration

A selective measurement is understood to be when the emf (or voltage) at the input of an arbitrary (loop) antenna is measured by a selective (micro-) voltmeter that allows selection of a single frequency from the spectrum radiated by an STA excited by an achromatic source. The errors for a peak detection (δ_{sp}) and rms (δ_{se}) are as follows:

$$\delta_{sp} = \sum_{n=2}^{\infty} \frac{I_n}{I_1} \tag{5.20}$$

and

$$\delta_{se} = \frac{1}{2} \sum_{n=2}^{\infty} \left(\frac{I_n}{I_1} \right)^2 = \frac{h^2}{2} \tag{5.21}$$

where: h – clear-factor:

$$h = \sqrt{\sum_{n=2}^{\infty} \left(\frac{I_n}{I_1}\right)^2} \qquad (5.22)$$

b. Wideband meter calibration

The frequency independence of the response of a wideband H-field probe is usually achieved by connection of the output emf (voltage) of a loop antenna to a voltmeter through a low-pass filter whose frequency response is proportional to 1/f. As a result, the response of the probe is flat in a required frequency band. Outside the band, at higher frequencies, the response may be an arbitrary one. However, in order to simplify our discussion, we will assume that the frequency response of the probe is flat over the whole frequency spectrum (this assumption does not introduce a gross error because usually only several prime harmonics play a role in the estimation). The measuring errors for peak detectors (δ_{wp}) and rms (δ_{we}) may be presented in the form:

$$\delta_{wp} = \sum_{n=2}^{\infty} \frac{\sqrt{1 + n^2 k^2 R_0^2}}{\sqrt{1 + k^2 R_0^2}} \qquad (5.23)$$

and

$$\delta_{we} = \frac{1}{2} \sum_{n=2}^{\infty} \frac{I_n^2 \left(1 + n^2 k^2 R_0^2\right)}{I_1^2 \left(1 + k^2 R_0^2\right)} \qquad (5.24)$$

c. Measurements with an SRA

As has already been presented, the output emf (voltage) of a standard receiving loop antenna is measured using a wideband voltmeter. Because of the proportionality of the effective height of a loop antenna to frequency, the sensitivity of such a system increases with frequency. Although at much higher frequencies the relation is much more complex, taking into account the role of the first harmonics, we will assume that the relation is valid over the whole frequency spectrum. Thus, the errors δ_{ap} and δ_{ae} are:

$$\delta_{ap} = \sum_{n=2}^{\infty} \frac{n H_n}{H_1} \qquad (5.25)$$

and

$$\delta_{ae} = \frac{1}{2} \sum_{n=2}^{\infty} \frac{n^2 H_n^2}{H_1^2} \qquad (5.26)$$

Table 5.1   Estimated values of errors caused by achromatic excitation of an STA (in %)

	δ_{sp}	δ_{se}	δ_{wp}	δ_{we}	δ_{ap}	δ_{ae}
$kR_0 = 0$	11,1	0,55	11,1	0,55	23,4	2,5
$kR_0 = 1$	11,1	0,55	18,5	1,27	39,5	5,2

In order to illustrate levels of possible errors and their dependence on the type of antenna used and type of detection, estimations for $I_n = I_1 \, 10^{n-1}$ are presented in Table 5.1. Although $h > 10\%$ is rarely met, it may happen, and the data in the table are for illustration only. The aim of this presentation is to focus attention on a factor that is rarely taken into account and is important both in EMF standards and in exposure systems.

5.3   An EMF standard with Helmholtz coils

Contrary to previously discussed transmitting antenna standards, the Helmholtz coil contains two coaxially placed coils, whose windings are connected in series. The solution is applied in order to enlarge an area in which the H-field is more uniform compared to a single coil (loop antenna). For stationary conditions and negligible cross section surface area of the coil windings in relation to the surface area of a coil, in the plane perpendicular to the symmetry axis of the two coils (antennas) the H-field components of the two antennas are similar to that given by (2.37) for negligible values of kR_0. We rewrite them in the form:

$$|H_{AV}|_1 = \frac{n_1 \, I_1 \, r_1^2}{2 \, D_1^3} \tag{5.27}$$

and

$$|H_{AV}|_2 = \frac{n_2 \, I_2 \, r_2^2}{2 \, D_2^3} \tag{5.28}$$

where: n_1 and n_2 – number of turns of the first coil and the second one, respectively,
r_1 and r_2 – radii of the coils,
I_1 and I_2 – current in the coils,
D_1 and D_2 – distance from a coil to a point of observation:

$$D_1 = \sqrt{r_1^2 + y^2 + (a + x)^2} \tag{5.29}$$

and

$$D_2 = \sqrt{r_2^2 + y^2 + (a - x)^2} \tag{5.30}$$

Other denotations are given in Figure 5.4.

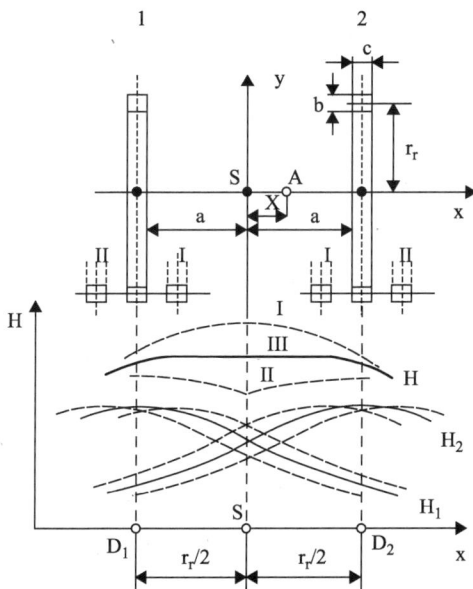

Figure 5.4 A Helmholtz coil and the H-field distribution inside it

In the majority of cases of practical importance, the two coils are identical in terms of their size and number of turns. Thus, for assumed static conditions [fulfilled condition of (5.13)], the currents in the two coils are identical as to their phase and amplitude. This leads to further similarity of the formulas expressing the H-field generated by every one of the coils [(5.27) and (5.28)]. Of course, an assumption of stationary fields indicates that that a set is designated for work at lower frequencies rather than higher ones. However, Helmholtz coils usually work at low frequencies, as mentioned. The resultant H-field generated by the two coils is a superposition of field H_1 and H_2:

$$|H_{AV}| = |H_{AV}|_1 + |H_{AV}|_2 = \frac{n I r^2}{2}[(D_1^{-3})^2 + (D_2^{-3})^2]^{1/2} = \frac{n I r^2}{2}f(x) \qquad (5.31)$$

In (5.31), f(x) illustrates spatial variations of the H-field distribution around the two coils. Figure 5.4 indicates the resultant field in the cross section of the coils when the coils are at an optimal distance (maximal field homogeneity), when the coils are located in position I (a maximum of the field in the center of the system) and in position II (a local minimum in the center).

In order to show a difference in the field homogeneity, Figure 5.5 presents H-field lines in the cross section of a single coil (left) and Helmholtz coils (right) when the distance between the coils is $2a = r$.

As may be concluded from Figure 5.5, the most homogeneous H-field exists only in the center of the single loop. In the case of Helmholtz coils, the

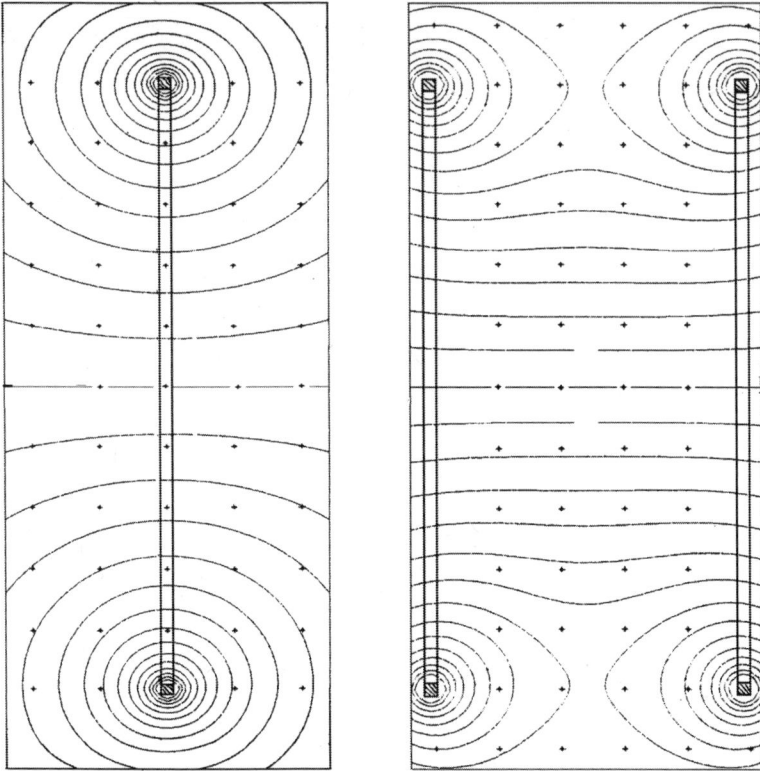

Figure 5.5   Distribution of the H-field lines in the center of a single coil (left) and Helmholtz coils (right)

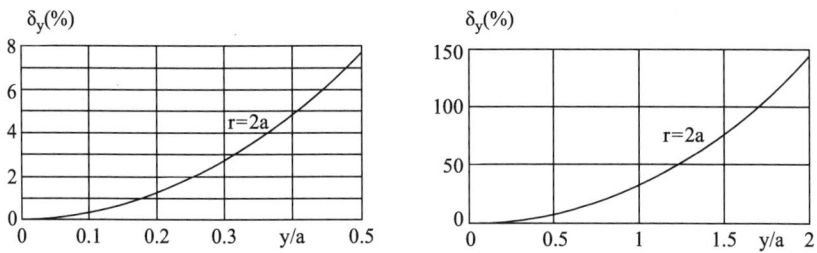

Figure 5.6   Changes of H-field intensity in the cross section of the Helmholtz coils

homogeneity is much better, especially in the plane equally separated from both coils, i.e., for $x = 0$. For this case we have:

$$H_{AV} = \frac{n\,I\,r^2}{D^3} \tag{5.32}$$

where:

$$D = \sqrt{r^2 + a^2 + y^2}$$

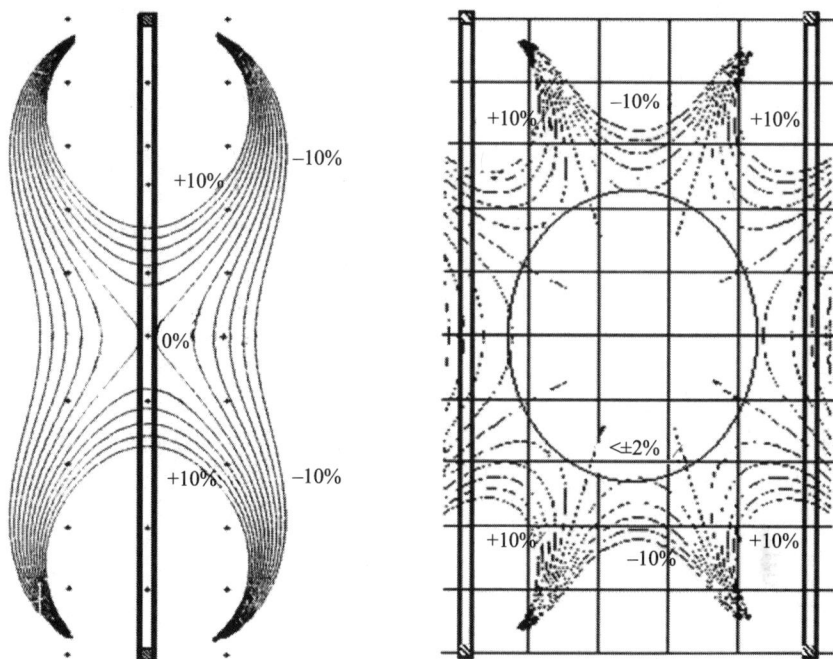

Figure 5.7 Dispersion in the H-field distribution in the case of a single loop (left) and a Helmholtz coil (right)

Figure 5.8 Multiloop and multi-turn systems designed by the authors: a set for EMC testing (left), an H-field probe measurements within a Helmholtz coils (center), and an LF H-field exposure system (right)

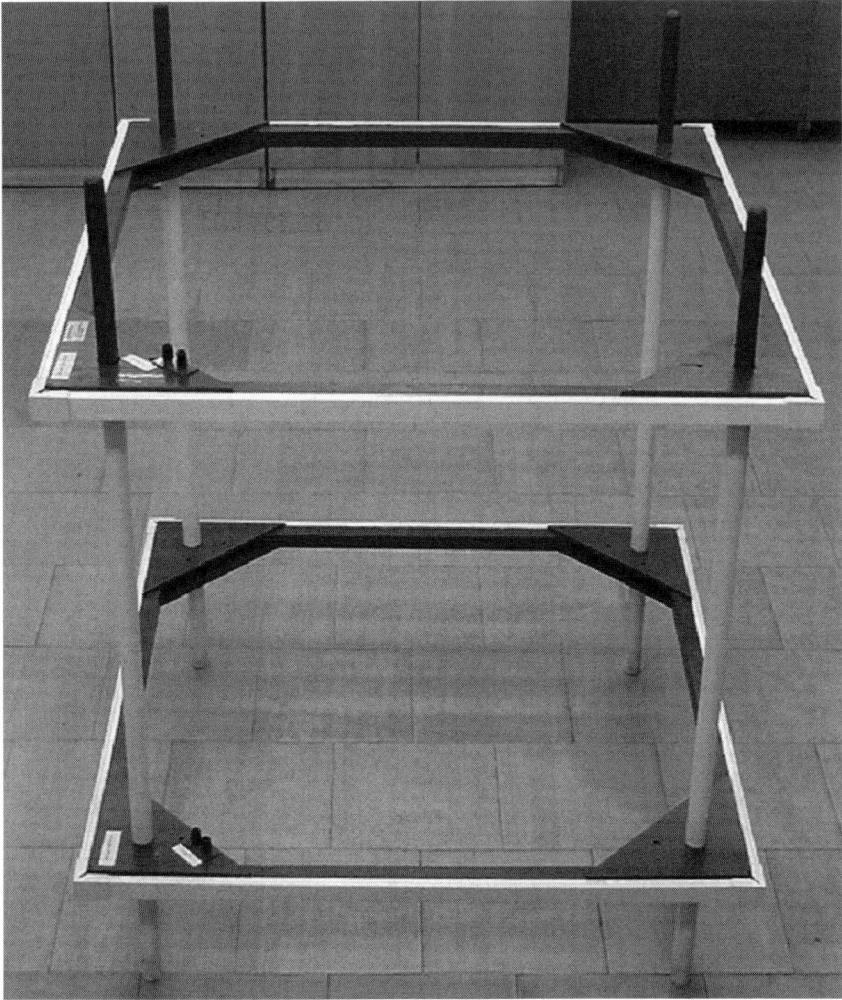

Figure 5.9   A large Helmholtz coil for EMC investigations

As may be seen, (5.32) is identical to (2.37), with the exception of an omitted factor (kR_0), which is fully acceptable for conditions assumed in the discussion and for frequencies below 5–10 MHz.

The discussion presented has confirmed that, in the case of the Helmholtz coils, it is a possible to reach an H-field of twice as great intensity as with a single loop. However, if the number of turns in the single loop were doubled, the effect would be identical. The main advantage of the Helmholtz coils is that a large volume of homogeneous H-field can be generated compared to that of a single loop. Of course, it is possible to have a large volume of homogeneous H-field in the case of any source. However, this is at distances much larger than the sizes of the antennas used. The case is only of theoretical importance, as the H-field intensity in this area is much smaller compared to that in the vicinity of a single loop or Helmholtz coils, assuming the same power exciting any of the sources.

In order to present a precise representation of the H-field homogeneity within the Helmholtz coils, Figure 5.6 presents results of calculations of the H-field dispersion (δ_y) in relation to that in the center of the system as a function of y/a.

The H-field homogeneity around a single loop is much less than that of the Helmholtz coils. Figure 5.7 shows the percentage changes of the H-field in a single loop and in Helmholtz coils in relation to their geometrical centers. The maps of the H-field distribution indicate areas in which the H-field differs by 10% from that at the center of considered system. In the case of the Helmholtz coils, an area of 2% dispersion is additionally indicated.

Apart from the presented system of two coils, other approaches are in use that may include several coils, a single long spatial coil, and others. Examples of different constructions of such systems, designed and applied by the authors, are shown in Figures 5.8 and 5.9. Regardless of the advantages of Helmholtz coils and the other similar solutions mentioned, their use is limited rather to lower frequencies and in the role of exposure systems rather than as primary standards. The reason for this is that it is much more difficult to estimate the accuracy of such a system.

Chapter 6

Accuracy analysis of EMF standards with a segment of a transmission line

As has already been mentioned, an EMF standard with a segment of a transmission line, working in conditions of full matching in both its sides, has many advantages compared to sets radiating EM energy into space. The main advantages include a simple design, a possibility to generate quite strong EMF with relatively low exciting power, insensitivity to external fields and very limited energy radiated outside of a set, an ability to work in a wide frequency range, and simple and frequency-independent relationship between power (voltage) exciting a set and EMF inside it. The most important disadvantages include very limited space inside a set and strong couplings between a line and an OUT. The most popular and the most widely applied solution consists of a segment of strip line with basic mode TEM; this is usually called a TEM cell.

Apart from the disadvantages mentioned, the TEM cell is applied as a primary EMF standard and in many studies as an exposure system. For instance:

- an indispensable tool in experimental research on near-field EMF measurements,
- testing and calibration of a variety of EMF detectors, indicators, probes, meters, and similar devices,
- calibration of transfer and secondary EMF standards,
- studies of properties of living and nonliving matter exposed to EMF,
- susceptibility testing of electric and electronic devices and systems,
- a basic tool in biomedical studies *in vivo* and *in vitro*,
- measurements of EM radiation generated by animals and technical devices,
- radiation measurements from crushed rocks,
- tests with flammable and explosive materials, and many others.

As was shown in section 3.4, a variety of implementations are possible, from the simplest plate capacitor up to sophisticated versions of different types of cells and waveguides. The basic solution here is a TEM cell, which ensures the highest accuracy. Thus, our attention will be focused on the cell. However, the considerations presented are valid for other solutions of standards and exposure systems as well.

6.1   EMF in a strip line

A block diagram of the EMF standard with a part of the strip transmission line is shown in Figure 3.18. An EMF distribution in a cross section of the symmetric strip transmission line is shown in Figure 6.1.

The line (TEM cell) shown in Figure 6.1 is an open-sided one. Sometimes, in order to better separate the environment inside the cell from the outside, sidewalls are added, as shown in Figure 3.17. This separation is especially necessary when high-intensity EMF is to be generated in the cell, for instance, in EMC tests or for near-field EMF probe calibration, in order to limit radiated EMI. Apart from some different EMF distribution in the two solutions, their working conditions are almost identical.

In conditions of a cell well matched to its excitation source and load, and neglecting energy losses in the cell itself, the E-field intensity in the area, where the field could be assumed as homogeneous, is given by (3.12). We rewrite that formula in modified form:

$$E = \frac{2\sqrt{P_e Z_c}}{2D - t} \exp(-jkz) \tag{6.1}$$

where:

z – current length of the cell,
$2D$ – distance between sidewalls of the cell,
t – thickness of the internal conductor (see Figure 6.1)

Similar modifications may be introduced to (3.13). Neglecting the phase along the cell, we may write:

$$H = \frac{E}{Z} = \frac{2}{2D - t} \sqrt{\frac{P_e}{Z}} q \tag{6.2}$$

where: q – coefficient of the wave impedance reduction in the cell:

$$q = \sqrt{\frac{Z_c}{Z}} \tag{6.3}$$

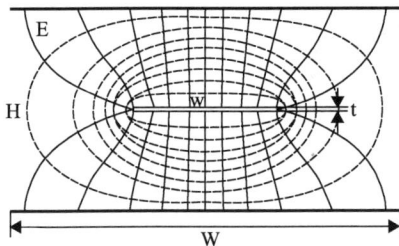

Figure 6.1   Cross section of a symmetric strip transmission line

For instance: For commonly used cells, with the wave impedance $Z_c = 50\ \Omega$, the factor $q = 0.364$.

As may be seen from the above equations, the EMF intensity inside the cell is, in the first approximation, a function of two variables, i.e., a power (voltage) exciting the cell and its geometrical size. The formulas are valid for the basic mode in the cell, i.e., the TEM mode. This assumption limits the cell use to the frequency range in which the mode is dominant and the presence of higher modes (which may be a result of intentional or unintentional EMF distribution deformations in the cell) is negligible. Of course, the presence of the higher modes is an important factor that limits the accuracy of the cell as an EMF standard, and the possibility of their presence, and the role played by them, should always be checked before and during measurements.

6.2   Wave impedance of a TEM cell

A wave impedance of any transmission line with a basic mode TEM is given by:

$$Z_c = \frac{1}{cC} \tag{6.4}$$

where: c – velocity of light in the vacuum:

$$c = \left(\varepsilon_0 \mu_0\right)^{-1/2} \tag{6.5}$$

C – unitary capacitance of the line.

In the case of a TEM cell, of which a cross section is shown in Figure 6.1, the capacitance C is the sum of capacitances of two flat capacitors created by the walls and the center conductor (C_p) and side capacitances (C_d), which represent dispersed fields at the sides of the cell and the sidewalls (if present) of the cell (Figure 6.2). Thus:

$$C = 2\left(C_p + 2C_d\right) \tag{6.6}$$

In the first approximation, especially for lines with relatively large C_p, the influence of the dispersed fields and side capacitances is negligible. A rigorous solution of

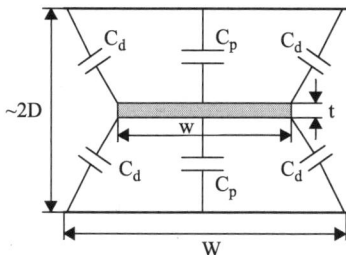

Figure 6.2   Distribution of unitary capacitance components in the cross section of a TEM cell

the problem of the capacitances leads to the necessity of finding the potential Φ that would fulfill the Laplace equation:

$$\nabla^2 \Phi = 0 \tag{6.7}$$

The Laplace equation must be fulfilled in any cross section of the cell. The problem has already been solved for the line sizes (as in Figures 6.1 and 6.2), fulfilling the condition $W \gg w$, and is expressed by:

$$C_d \approx \frac{\varepsilon}{\pi} \left\{ \frac{2}{1 - t/2D} \ln \left[\frac{1}{1 - t/2D} + 1 \right] - \left[\ln \frac{1}{(1 - t/2D)^2} - 1 \right] \right\} \tag{6.8}$$

For the line fulfilling the condition $t \ll 2D$, (6.8) takes the form:

$$C_d \cong \frac{2\varepsilon}{\pi} \ln 2 \tag{6.9}$$

and C_p is:

$$C_p = \frac{2\varepsilon w}{2D - t} \tag{6.10}$$

Substituting (6.9) and (6.10) into (6.6) and then into (6.4), we get:

$$Z_c = \frac{15\pi^2}{\sqrt{\varepsilon_r}} \left(\frac{\pi w}{4D} + \ln 2 \right)^{-1} \tag{6.11}$$

A plot of C_d, normalized in relation to ε, as a function of $t/2D$ is shown in Figure 6.3. From the point of view of the present discussion, it is important to draw attention to the dependence of the line wave impedance on the capacitance. The capacitance may be changed by an object immersed in the line and, as a result, the wave impedance will be affected and a mismatch will appear. The case is discussed in section 6.4.

The equations introduced above are valid for the basic mode (TEM). As a result, frequency does not appear in the formulas. However, frequency limitations exist and they are a result of the presence of a corner wavelength λ_b, below which the higher modes exist. The wavelength may be estimated as follows:

$$\lambda_b < w/2 \quad \text{and} \quad \lambda_b < D/2 \tag{6.12}$$

There is a possibility of eliminating E_{mn} and H_{mn} higher mode type waves by application of appropriate slots in the center conductor (Figure 6.4), which attenuates the transversal current component, or by use of absorbing material [45]; however, these solutions are of limited practical importance. The only effective method of frequency range widening in a TEM cell, with preservation of the basic mode presence, is a reduction of the cell size [as given by (6.12)]. However,

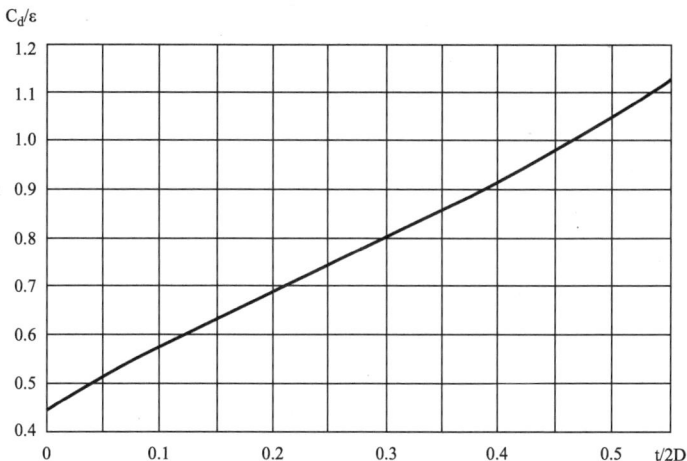

Figure 6.3 C_d/ε as a function of t/2D

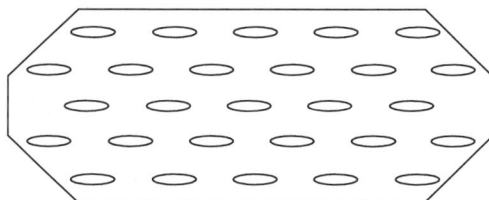

Figure 6.4   Slots limiting the tangential current component in the center conductor of a TEM cell

construction of a TEM cell working at frequencies above 1 GHz may create technological problems. Moreover, the presence of higher modes in the cell may result from the presence of one or more objects investigated in the cell. The latter may, in a first approximation, be neglected when a small-size probe is calibrated in the cell; however, it may play an essential role when the cell is applied as an exposure system. A possibility of using a TEM cell with the presence of higher modes as an exposure system was presented in section 3.5.2.

Equation (6.12), and similar relationships valid for any type of waveguides, are introduced for "infinitely long" devices, and they do not take into account their connection to other elements of the system. As has already been presented in section 3.4.2, the TEM cell is matched to other parts of an HF or microwave track by two transformers. The role of the transformers is twofold: a matching wave impedance of a TEM cell to other parts (the wave impedance of the cell may be different as a standard impedance of the track, usually 50 Ω) and a change of cross section geometry in order to adapt it to coaxial connectors in the track. It should be noted that the transformers cause EMF disturbances that may be a source of higher modes. In order to limit undesirable effects caused by the transformers, different lengths of the external and internal parts of the transformer are used compared with the simplest triangle shape. It is usually assumed that the length of the transformer l_t should fulfill the following condition:

$$l_t < 0.1\, \lambda_{max} \tag{6.13}$$

where: λ_{max} – the maximal wavelength at which the cell is to work.

This condition is usually not fulfilled because at lower frequencies it would require the construction of long devices that may limit their application at higher frequencies. This leads to a practical approach in which it is usually assumed that $l_t \approx l_c/3$ (where: l_c – total length of the cell).

Regardless of the above considerations, the EMF distribution in the areas near the transformers is disturbed, which provides additional limitations on the operational part of the cell to its central area. The limitation is not very troublesome when small EMF antennas or sensors are being calibrated in the cell; however, it should be taken into account when the device is applied as an exposure system and larger objects are exposed in it. In solutions presented in section 3.5.2, usually a single transformer appears. However, the solutions do not ensure the required accuracy, and they are usually applied as secondary standards and exposure systems, where accuracy is less rigorously required.

6.3   Accuracy of the line accomplishment and its role in line accuracy limitation

The accuracy of the line accomplishment is the simplest to analyze and in which to eliminate sources of errors. The presence of inaccurate construction causes a mismatch of the line that is easily measurable. Taking (6.11) as a starting point for the analysis, the mean square error of the wave impedance calculation δ_z may be defined in the form:

$$\delta_z = \sqrt{\delta_w^2 + \delta_D^2} \tag{6.14}$$

where: δ_w and δ_D – error of the linear size assignment (in this case accuracy of the center conductor wide measurement [w] and that of the distance between sidewalls of the line [D]).

Assuming, as previously, that an error in the linear size measurement does not exceed 1%, and δ_w and δ_D are of a similar level, the error δ_z should not exceed 1.4%. This error is mentioned here only for formal correctness of the discussion. It illustrates mismatching of the cell with no load and may be immediately eliminated by using a matching network at the input of the cell. On the other hand, regardless of the accuracy of the device, a mismatch is always caused when an OUT is placed into the cell. However, the best matching is unable to correct EMF distribution in the cell and the distribution is a function of the line matching and the presence of standing waves in the line. We may refer here to the case of a plate capacitor (see section 3.4.1) where reflection factor $\Gamma \approx 1$, but its size in relation to the wavelength is small. The relation between the reflection factor and electrical size of the line and the reflection factor will be discussed in detail in the next chapter.

A much more important consideration here is the noncentric placement of the center conductor in relation to the sidewalls of the line. Although a cell with noncentric placement of the conductor is sometimes used (usually to increase the usable volume of the cell), (3.11) is valid and shows the direct dependence between the distance D and the field intensity in the space. Let's consider the issue in terms of its role in cell mismatching. For nonsymmetrical placement of the center conductor, (6.6) may be rewritten in the form:

$$C_n = C_p' + C_p'' + 2(C_d' + C_d'')$$ (6.15)

where: C_n – unit capacitance of nonsymmetrical line,
$\quad$ C_p' and C_d' – capacitances between center conductor and the wall at distance
$\quad\quad$ $D + \delta$,
$\quad$ C_p'' and C_d'' – capacitances between center conductor and the wall at distance
$\quad\quad$ $D - \delta$,
$\quad$ δ – eccentricity of the center conductor in relation to its distance to sidewalls equal to D.

If we calculate the relation of the noncentric line wave impedance Z_c' to the symmetric impedance Z_c we may write:

$$\frac{Z_c'}{Z_c} = \frac{C}{C_n} = \frac{\left(1 - \dfrac{\delta}{2D} - \dfrac{t}{2D}\right)\left(1 + \dfrac{\delta}{2D} - \dfrac{t}{2D}\right)}{\left(1 + \dfrac{t}{2D}\right)^2}$$ (6.16)

In the case of a line that fulfils the relation $t \ll D$, (6.16) may be reduced to the form:

$$\frac{Z_c'}{Z_c} = 1 - \left(\frac{\delta}{2D}\right)^2$$ (6.17)

As above, the mismatching of the line is not of primary importance. The presence of mismatching in an empty line is an illustration of the accuracy of its assembly and may be reduced by an introduction of mechanical corrections or by the use of a matching network. The most important consideration here is asymmetry of the E-field in the two parts of the line. This asymmetry may be expressed in the form:

$$\frac{\Delta E}{E} \cong \pm \frac{\delta}{D}$$ (6.18)

6.4   Disturbances of the wave impedance

The sources of line mismatching, discussed above, involved inappropriate line design and/or inaccurate assembling. When considering a line as an EMF standard, it is possible to assume that the necessary corrections were introduced and that the actual reflection factor at the line's input fulfills our requirements. An OUT is introduced to a line that meets the above criteria, which introduces reflections and line mismatching proportional to the object's size and the material from which it is made (i.e., its electrical parameters). As a result, the wave impedance of the line across its cross section may be different. This case is illustrated in Figure 6.5. An OUT of length, l_z, is immersed in a two-wire line of total length l_c. As a result, the wave impedance of the line within the object may vary, as shown in Figure 6.5.

The mean value of the wave impedance within the disturbed area is Z'_{cm}, while the current impedance $Z'_c(z)$ is given by:

$$Z'_{cm} = \frac{1}{l_z} \int_0^{l_z} Z'_c(z)\,dz \tag{6.19}$$

As a result, the relative wave impedance of the line is changed as $\Delta Z_c/Z_c$:

$$\frac{\Delta Z_c}{Z_c} = 1 - \frac{Z'_{cm}}{Z_c} \tag{6.20}$$

A change of the local wave impedance in the line is not a direct reason for calibration errors. The errors result from nonuniform voltage distributions within the line, as a result of standing waves present in it due to line mismatching (see next chapter).

The main application a TEM cell as an EMF standard is its use in EMF probe, sensor, and antenna calibration. Such a device, placed in a line, may be represented

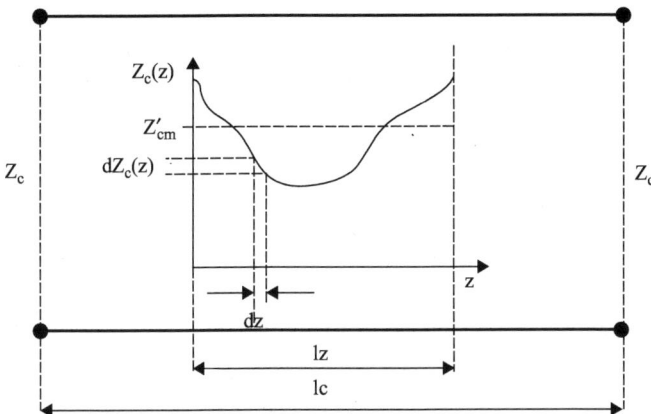

Figure 6.5   Two-wire line and its wave impedance in the area of an OUT

by a conducting diaphragm. Let's consider, for illustration, the role played by a diaphragm representing a calibrated device. A thin and perfectly conducting diaphragm is immersed between infinitely long, two-plate lines perpendicularly to the line and in its center, as shown in Figure 6.6.

The presence of the diaphragm is equivalent to a local increase of the unit capacitance of the line (Figure 6.7). The susceptance B, representing the diaphragm, is given by:

$$B = \frac{4kDa}{\pi w} \ln \cosec \left(\frac{\pi d}{D} + \frac{L\alpha_1^2}{1 + L\alpha_2^2} \right) Z_c \qquad (6.21)$$

where: $L = \dfrac{4\pi}{\gamma} - 1$

$$\gamma = \sqrt{\left(\frac{2\pi}{D} \right)^2 - k^2}$$

$$\alpha_1 = \cos \frac{\pi w}{D}$$

$$\alpha_2 = \sin \frac{2\pi d}{b}$$

$$k = \frac{2\pi}{\lambda}$$

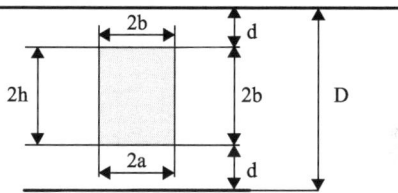

Figure 6.6   A diaphragm in a line

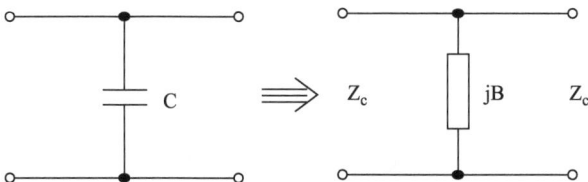

Figure 6.7   Equivalent circuit of a diaphragm in a line

The susceptance causes an increase in the voltage standing wave ratio (VSWR), which for $|B| \ll Z_c$ is given by:

$$\text{VSWR} \approx \frac{1}{2}\left(\frac{B}{Z_c}\right)^2 \tag{6.22}$$

Table 6.1 presents a comparison of estimations and measurements of the VSWR at the input of a symmetric TEM cell when an electric field probe was immersed in it. As may be seen from the table, the estimated values are smaller than the measured ones. The difference may be a result of the following:

- In the calculations a probe of finite thickness was represented by a thin diaphragm.
- The VSWR, at frequency 10 MHz, was measured by way of the line capacitance change measurement, which is loaded by relatively large errors; at a frequency of 300 MHz the measurement was performed using the Zg-diagraph, but without the probe in the cell VSWR was around 1.05.

Now we will show the results of several estimations performed for an electric dipole antenna of size $2a \times 2h$ (Figure 6.8) and a square loop antenna of size $2b \times 2b$ (Figure 6.9) immersed in the center of the line, as shown in Figure 6.6.

The calculated relationship between the normalized susceptance and the modulus of the reflection factor Γ, for an object placed in the center of the cell, is shown in Figure 6.10. As an example, the VSWR measurement results for a TEM cell, one of the models designed and tested by the authors, of total length $l_c = 1.3$ m and $D = 0.25$ m, are shown in Figure 6.11. A remarkable increase in VSWR may be seen at 375 and 475 MHz; this excludes use of the cell around these frequencies.

Figure 6.11 presents measurements of reflections in an empty cell. It may be seen that certain frequencies exhibit strong reflections. Papers are available reporting EMC or biomedical investigations using a cell of length exceeding λ. In the latter case, particularly, the cell contains as many objects as it is possible to immerse into the cell. In light of the curves in Figure 6.11, it is impossible to say what is a matching of such a cell and what is a field distribution inside it, whereas the absorbed energy in separate objects is estimated as energy lost in the cell divided by the number of objects. The issue is further discussed in section 6.8.2; however, this is a good place to warn against possible errors and an application of such an approach.

Table 6.1   Results of VSWR estimations and measurements

	Estimations			Measurements		
2h/D	0.88	0.56	0.32	0.88	0.56	0.32
10 MHz	1.001	1.0003	1.00004	1.003	1.001	1.001
300 MHz	1.03	1.007	1.001	1.15	1.10	1.05

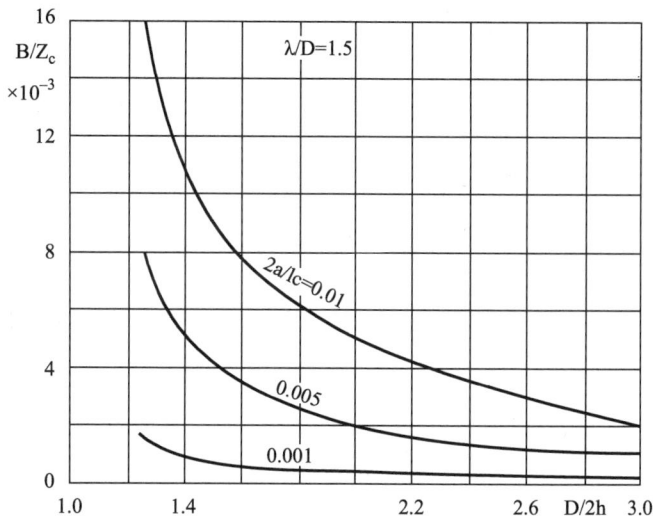

Figure 6.8   Normalized susceptance B/Z_c of a diaphragm of size $2a \times 2h$, versus $D/2h$

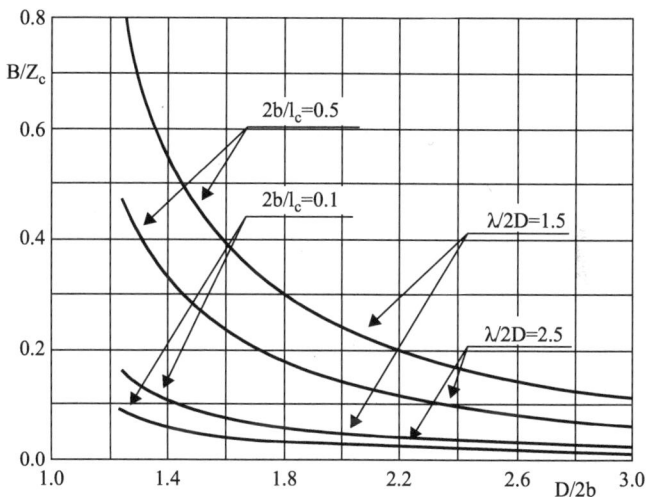

Figure 6.9   Normalized susceptance B/Z_c of a diaphragm of size $2b \times 2b$, versus $D/2h$

The present discussion and results of calculations may make it possible to make an optimal choice of a TEM cell for an OUT investigated in it, and an estimation of reflections caused by the object and its role in any inaccuracy of the performed measurements (calibrations). Although these considerations were made in regard to relatively small objects being calibrated in the cell (i.e., EMF probes),

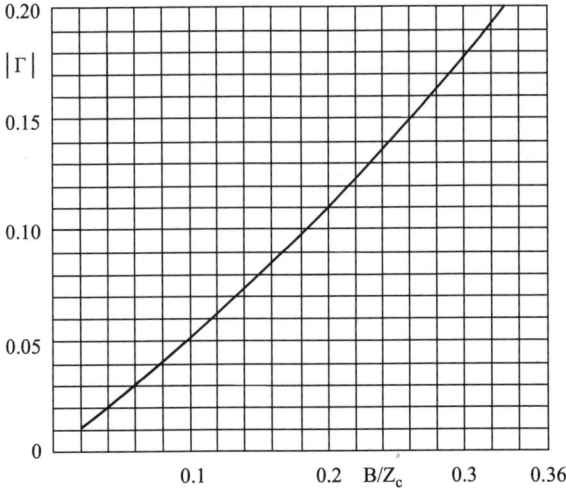

Figure 6.10   Relationship between normalized susceptance and Γ

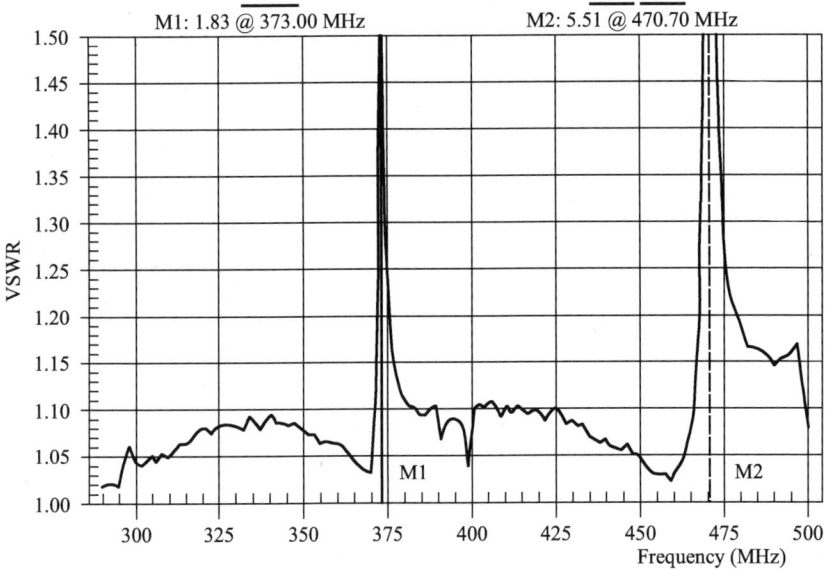

*Figure 6.11   Measured VSWR of a symmetric TEM cell of size $l_c = 1.3$ m,
            $D = 0.25$ m*

a similar approach should be taken when the cell is applied as an exposure system and the objects are much larger. As an example, Figure 6.12 shows an EMF indicator tested in the cell. The indicator consists of a metallic box containing a measuring system and a whip antenna. The top of the antenna is only 1 cm from the

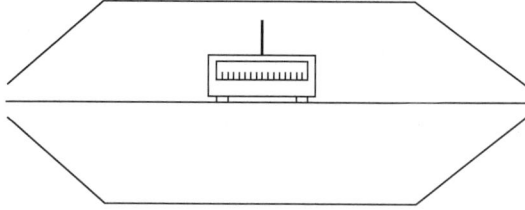

Figure 6.12   An EMF indicator in a TEM cell

sidewall of the cell. The indicator in the cell causes VSWR $\approx$ 2 at frequency 50 MHz, and the VSWR increases with increasing frequency. This is to remind us once more that it is possible to compensate reflected waves by matching the cell loaded with an OUT to the wave impedance of the whole system. However, this does not change any EMF distribution inhomogenities within the cell. The problem of mismatching due to the presence of an OUT in the cell is of primary importance when the cell is applied as an exposure system and almost the whole volume of the cell is occupied by the OUT or OUTs.

6.5   Line excitation measurement accuracy

As was shown above, as a result of inaccurate line preparation and the presence in it of any material objects, the matching of the line is not ideal and the value of the mismatch mainly depends upon the size and electric properties of the object. As may be concluded from (3.11) and (3.12), the accuracy of the electric and magnetic field assignment in a transmission line is proportional to measurement accuracy of the power delivered to the line or the voltage between its conductors. However, the line's excitation measurements are performed at the input and/or at the output of the line usually when the OUT is immersed in the line center. Thus, due to a presence of standing waves within the line, an effective voltage in the center is a function of the line matching, and it may be different than the values measured at the ends. Even if the line (including an OUT) is well matched, the voltage distribution along the line remains unchanged, and it is a function of standing waves in the line. Now we will consider the relationship between the line matching accuracy and the reflection factor, Γ. Of course, for $\Gamma = 0$, any excitation measurement methods are equivalent to the accuracy to the class of excitation meters applied. In real cases, for $\Gamma \neq 0$, a difference between measured excitation and the EMF in the place of the OUT may exist. This leads to an error in the EMF assignment in place of the OUT. The error is of a systematic character, but different for different objects, frequencies, line lengths, etc. It is possible to measure reflections caused by an OUT using time domain reflectometry – even reflections in the area in which the OUT appears. However, the procedures are troublesome, and it is not always possible to apply them, especially when larger objects are involved. Below, we will consider two cases of allowable reflection levels when the difference between measured excitation and voltage in the center of the cell is $\pm 2.5\%$ and $\pm 5\%$, i.e.,

$$\left|\frac{\Delta V}{V}\right| = \pm 0.025 \text{ and } \left|\frac{\Delta V}{V}\right| = \pm 0.05 \tag{6.23}$$

There are many methods of measuring cell excitation; we will take into account three of them: voltage measurement at the input of the cell, incident voltage measurement at the input of the cell, and simultaneous voltage measurement at the input and at the output of the cell. In our discussion, we will assume that the wave impedance Z'_c of the cell is identical in each of its cross sections. This assumption will maximize the presented errors because, usually, the OUT is placed in the center of the cell and the mismatching appears only in the part of the cell where the OUT is placed. However, this is only an example of one way to approach the problem and of the role of mismatching and the limitations in accuracy caused by it. Similar estimation should be performed for any particular case of TEM cell use. A schematic diagram of the analyzed case is shown in Figure 6.13.

If multiple reflections are neglected, the voltage $V(z)$ along the line is given by:

$$V(z) = e_g \frac{Z'_c}{Z'_c + Z_g} \frac{\exp(-jkz) + \Gamma_2 \exp[-jk(2l_c - z)]}{1 - \Gamma_1 \Gamma_2 \exp(-j2kl_c)} \tag{6.24}$$

where: $\Gamma_1 = \dfrac{Z_g - Z'_c}{Z_g + Z'_c}$

$\Gamma_2 = \dfrac{Z_l - Z'_c}{Z_l + Z'_c}$

e_g – electromotive force of the generator,
Z_g – output impedance of the generator,
Z'_c – wave impedance of line with an OUT,
Z_l – loading impedance of the line.

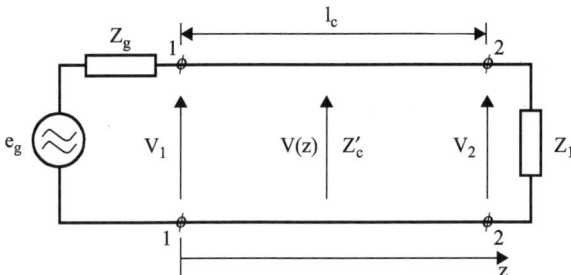

Figure 6.13   Schematic diagram of an analyzed model

For matching conditions, $Z_g = Z_l = Z_c = Z_0$, and it is equal to the assumed standard wave impedance, e.g., 50 Ω. In the case of mismatching, $Z_g = Z_l = Z_0$, and as a result, $\Gamma_1 = \Gamma_2$.

6.5.1   *Inaccuracy due to input voltage measurement*

We will consider two cases: first, when the line is not loaded, i.e., $Z_l \to \infty$ and, as a result, $\Gamma_2 \to \infty$; and second, when the line is loaded by loading impedance Z_1.

If the line is not loaded, (6.24) takes the form:

$$V(z) = -\frac{e_g}{\Gamma_1} \frac{Z'_c}{Z'_c + Z_g} \exp(-jkz) \tag{6.25}$$

If we refer the voltage at the center of the line ($V_{1c/2}$) to the voltage at its input (V_{1c}), we will have:

$$\left| \frac{V_{1c/2}}{V_{1c}} \right| = \exp(-jkl_c/2) \tag{6.26}$$

If we now use limits assumed by (6.23), we will have:

$$0.975 \leq \left| \frac{V_{1c/2}}{V_{1c}} \right| \leq 1.025 \text{ and } 0.95 \leq \left| \frac{V_{1c/2}}{V_{1c}} \right| \leq 1.05 \tag{6.27}$$

Comparing (6.26) and (6.27), we may conclude that (6.27) is fulfilled when $l_c \to 0$ or $\lambda \to \infty$. This case illustrates conditions that should be fulfilled by an unloaded line that, as discussed in section 3.4.1, represents a standard with a plate capacitor. However, the above discussion allows a relationship to be found between the required accuracy and acceptable length (sizes) of the capacitor.

In the case of a line loaded by Z_1, the relationship between voltage ($V_{1c/2}$) in the center and voltage (V_1) at the input is expressed by:

$$\left| \frac{V_{1c/2}}{V_1} \right| = \sqrt{\frac{1 + |\Gamma_2|^2 + 2|\Gamma_2|\cos kl_c}{1 + |\Gamma_2|^2 + 2|\Gamma_2|\cos 2kl_c}} \tag{6.28}$$

The equation is plotted in Figure 6.14 for conditions given by (6.27), i.e., for differences in the voltages at levels $\pm 2.5\%$ and $\pm 5\%$.

We can see that, if the line is short, the reflection factor may be quite large (see above case of a capacitor). However, if the line is of resonant size, and its length is a multiple of $\lambda/4$, the matching must be accurate if an error due to the discussed phenomenon is to be kept at an acceptable level.

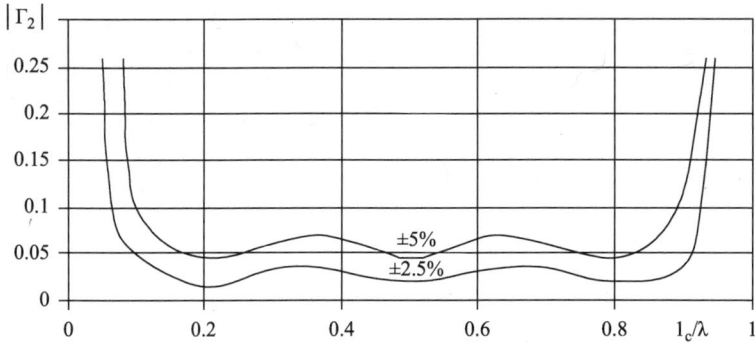

Figure 6.14   Allowable reflection factor as a function of the electric length of the line during input voltage measurement

6.5.2   Inaccuracy due to simultaneous input and output voltage measurement

If in (6.24) we substitute $z = l_c$ we will have a voltage V_2 at the line's end:

$$V_2 = V_1 \frac{\exp(-jkl_c) + \Gamma_2\exp(-jkl_c)}{1 + \Gamma_2\exp(-jkl_c)} \tag{6.29}$$

A modulus of the arithmetic mean of the voltages at the line input and output V_0 is:

$$|V_0| = \frac{|V_1| + |V_2|}{2} = \frac{|V_1|}{2}\left(1 + \sqrt{\frac{1 + |\Gamma_2|^2 + 2|\Gamma_2|}{1 + |\Gamma_2|^2 + 2|\Gamma_2|\cos kl_c}}\right) \tag{6.30}$$

Now we will refer the voltage V_0 to the voltage in the center of the line $V_{l_c/2}$:

$$\left|\frac{V_{l_c/2}}{V_0}\right| = \frac{2\sqrt{\dfrac{1 + |\Gamma_2|^2 + 2|\Gamma_2|\cos kl_c}{1 + |\Gamma_2|^2 + 2|\Gamma_2|\cos 2kl_c}}}{1 + \sqrt{\dfrac{1 + |\Gamma_2|^2 + 2|\Gamma_2|}{1 + |\Gamma_2|^2 + 2|\Gamma_2|\cos kl_c}}} \tag{6.31}$$

Results of calculations are plotted in Figure 6.15.

In this case, a line of length $l_c \approx \lambda/2$ or multiples of this length is the most sensitive to the presence of reflections. A TEM cell with voltage measurements at the input and at the output of the cell, constructed by the authors, is shown in

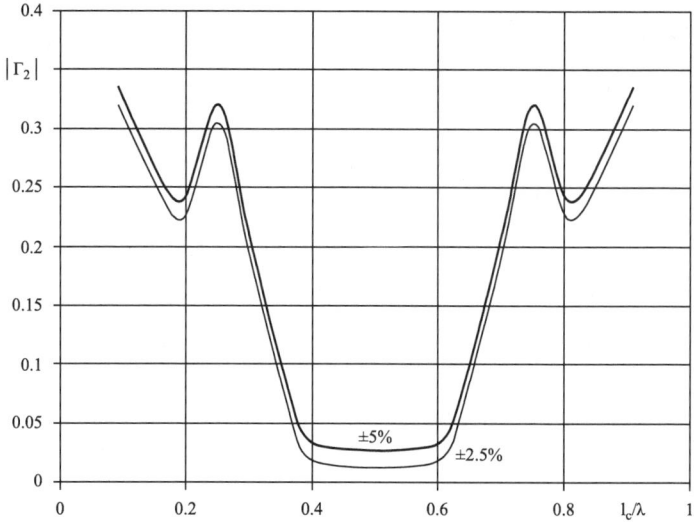

Figure 6.15   Allowable reflection factor as a function of the electric length of the line when input and output voltages are measured

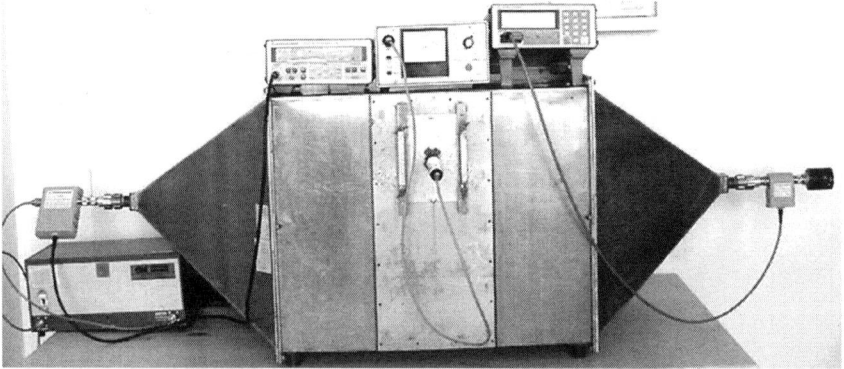

Figure 6.16   A TEM cell with voltage measurements at the both ends

Figure 6.16. Between two voltmeters, at the top of the cell, is placed a calibrated EMF meter. Its probe is placed inside the cell.

6.5.3   *Inaccuracy due to incident voltage measurement*

The voltage at the line input, V_1, may be expressed as the sum of an incident wave voltage V_1^+ and the voltage of a reflected wave V_1^-:

$$V_1 = V_1^+ + V_1^- \tag{6.32}$$

In order to simplify considerations, let's assume that $\Gamma_1 = 0$ and there are no multiple reflections in the line; then the relation of the reflected and incident waves may be written in the form:

$$\frac{V_1^-}{V_1^+} = \Gamma_2 \exp(-j2kl_c) \tag{6.33}$$

The voltage in the center of the line $V_{lc/2}$ is:

$$V_{l_c/2} = V_1^+ [\exp(-jkl_c/2) + \Gamma_2 \exp(-j3kl_c/2)] \tag{6.34}$$

and the ratio of voltage in the center to that of the incident wave is:

$$\left| \frac{V_{l_c/2}}{V_1^+} \right| = \sqrt{1 + |\Gamma_2|^2 + 2|\Gamma_2|\cos kl_c} \tag{6.35}$$

The ratio is plotted in Figure 6.17.

A set-up of a TEM cell excited from a generator through a reflectometer that allows incident wave voltage measurement, applied by the authors for probe calibration, is shown in Figure 6.18.

In our estimations, it was assumed that the wave impedance of the considered loaded line was constant along its length. As a result, only two reflections (Γ_1 and Γ_2) were taken into consideration. The presence of multiple reflections was neglected. These assumptions are not fully representative of situations that may occur in practice. Usually, a load distribution inside the line is nonuniform and may be arbitrary (as shown in Figure 6.5). However, regardless of any assumptions and simplifications, the presented analysis is focused upon a phenomenon that is usually not taken into account when a TEM cell, or similar solution, is used.

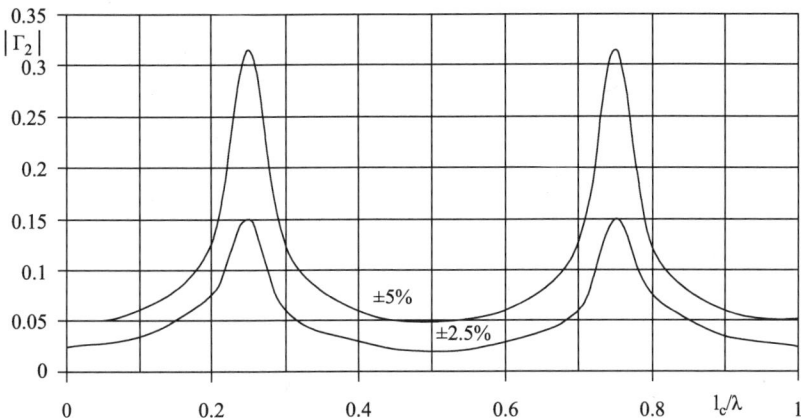

Figure 6.17   Allowable reflection factor as a function of the electric length of the line when incident voltage is measured

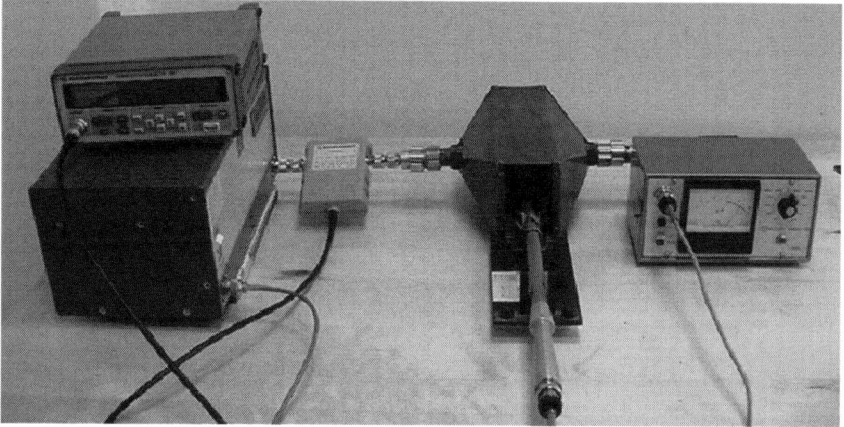

Figure 6.18 A set-up with measurement of the incident wave voltage

The analysis allows optimization of the excitation measurement and selection of a method of measurement that will ensure maximal accuracy. For instance:

- In the case of relatively short lines, i.e., for $l_c < 0.1\lambda$, the most convenient approach is voltage measurement at the input of the line. This results from the allowable arbitrary values of the reflection factor; the case includes a standard with the plate capacitor that may be presented as an "open ended" transmission line.
- In the case of longer lines, but shorter than a half wavelength, a voltage measurement at both ends of the line seems optimal.
- When the voltage of the incident wave is measured, the matching requirements are the most rigorous, and this approach is suggested for line applications as an EMF standard (where reflections due to a probe calibrated in the line can be omitted), but not in the case of an exposure system, where the presence of reflections is usually not controllable.

To repeat: These considerations as presented are based upon several rough assumptions and they have no direct application to any particular case. They are presented here only in order to focus attention on a problem that is usually ignored and to suggest its scale. Similar estimations, or experimental checking of mismatching in any experimental set-up, should be undertaken if accuracy of exposure (calibration) is of concern.

A word about simplifications: As shown in Figure 6.13, it was assumed that the line is directly connected to a generator and a matched load is connected directly at the line's end. In practice, other devices are connected at both ends of the line, not to mention the interconnecting cables (see Figure 3.16). At any point in the setup, there may appear a reflection, and any of these should be taken into account in the analysis instead of only the reflections at both ends of the line. Moreover, the real length of the track should be taken into account, including the track from a power source to the final load and not just the length of the line itself. This shows, on one hand, the approximations assumed in the presented analysis. On the other hand, it illustrates the impossibility of performing a strict analysis that would be of universal applicability. In any case, one very important conclusion resulting both from the presented considerations and from the experience of the authors is that the total length of interconnections should be as short as possible (see Figure 6.18).

6.6   Accuracy of the EMF standard with a segment of a transmission line

Although the considerations presented here are based mainly upon a TEM cell, they are valid for any other system in which a segment of a transmission line is used. Estimations presented in section 6.3 related to the accuracy of a line design and construction, and their influence upon the line wave impedance is very important from the point of view of precise manufacturing a line. However, mismatching resulting from mistakes in the line assembly may be measured with inaccuracy better than ±0.1%, and as a result, it' is possible to introduce any necessary corrections to the line and match it with required accuracy. This allows an assumption that this factor limiting the accuracy of the standard may be neglected. This statement may suggest that the considerations could be excluded from the presentation; however, they illustrate the role of mechanical problems, and they may be of some use when construction of a line is planned, especially when the line will be used as a primary EMF standard where the influence of a calibrated probe upon the line parameters is minimal. The main source of the reflections, line mismatching, and errors due to the reflections is an object exposed in the line.

If we assume that in the considered line $D \gg t$, $W \gg w$, and $l \ll \lambda$, then phase changes along the line can be neglected. If the line with undistorted monochromatic voltage V is measured with an arbitrary method, (6.1) may be rewritten in its simplest (and the most often used) form, identical to (3.11):

$$E = \frac{V}{D} \qquad (6.36)$$

The mean square error δ_E of the electric field assessment in the line will be given by:

$$\delta_E = \sqrt{\delta_1 + \delta_2 + \delta_3} \qquad (6.37)$$

where: δ_1 – excitation measurement error,
δ_2 – linear size measurement error,
δ_3 – error due to line mismatching.

Let's consider two cases:

1. An "empty" line, or a line loaded with an object that does not, effectively, cause reflections, VSWR < 1.01, i.e., a typical application as a primary EMF standard. In this case, one may assume that the dominant role is played here by the accuracy of the excitation measurement, and $\delta_E \approx \delta_1$. An excitation

measurement error depends mainly upon the class of measuring device applied. It may be assumed that, using medium-class meters, the error should not exceed $\pm 2\%$ and, as a result, the accuracy of the standard $\delta_E \approx \delta_1 \leq \pm 2\%$. At frequencies above 500 MHz, taking into account other factors is suggested as well; their presence degrades the accuracy of the standard.

2. A line loaded with an OUT giving a VSWR > 1.01. The presence of the reflections caused by the OUT has to be taken into consideration; in extreme cases, the role of reflections may dominate other factors. Estimations presented in section 6.5 lead to the conclusion that, due to mismatching caused by presence in the line an OUT and assumed method of line excitation measurement, the error of the EMF estimation in the area of the OUT may by arbitrary. If we assume that the reflections are not too large, VSWR < 1.2, the error due to the presence of mismatching should not exceed $\pm 4\%$, and as a result $\delta_E \leq \pm 5\%$. When an OUT occupies the majority of the space inside a line, which is the case for some EMC tests and for almost all exposures in biomedical investigations, the error may exceed 100% and the investigations are rather of a qualitative character than quantitative.

Although linear size measurement error is indicated in (6.37), in order to fulfill requirements, if we assume as previously that the error $\delta_2 \approx \pm 1\%$, its role may be neglected, especially at lower frequencies where mechanical design of a line may be more precise. One last comment: The accuracy of the standard magnetic field generation with a use of the line (TEM cell) is at the same level as for the electric field.

6.7   Homogeneity of EMF in a TEM cell

In the above discussion, it was assumed that the EMF in a line was homogeneous, undisturbed by the matching transformers and with the field disturbances caused by the OUT being negligible. The line is well matched in both its ends and it is fed from a monochromatic source. These conditions may be recognized as ideal ones, the aim of the standards designers. However, it is not possible to achieve all the conditions, especially when devices working at the highest frequencies are of concern. The latter situation leads to the necessity to reduce the line's size and, as a result, the volume at which the EMF is quite homogeneous decreases and relative size of the OUT increases.

The formulas introduced for a plate capacitor (3.11) and for a TEM cell [(3.12) and (6.36)] are true if the sizes of plates are much larger than the distance between them. The formulas do not take into account either EMF inhomogeneities in a real device or the field deformations caused by the OUT. The deformations were mentioned, however, and they were taken into account only in the previous chapter when accuracy of excitation was discussed. It could be said that, although very precise analytical and numerical tools are presently available, it is almost impossible to take into account all the EMF deformations in a line or their role in standard accuracy estimations, especially when larger objects are exposed in it. As was shown above, the applied assumptions and simplifications allow estimation of the accuracy in the case of small objects (antennas, EMF probes). To illustrate this, Figure 6.1 shows the E- and H-field distribution in the cross section of a symmetric strip line. The distribution in the case of a fully screened symmetric strip line is shown in Figure 6.19. Figure 6.20 presents the EMF inhomogeneities in the areas of matching transformers.

Results of estimations of the E-field inhomogeneity in the cross section of a symmetric strip line as a function of $y/D = 0.2-0.8$, for two different lines of given D/W and w/W ratio, are shown in Figure 6.21. The values of relative inhomogeneity presented show the ratio of E-field intensity at a point referred to the intensity in geometrical center of the line. Of course, this does not mean that the latter is precisely given by (6.36).

Field distributions presented in these figures illustrate the EMF inhomogeneity in a line without any object in it. This confirms that, even in an empty line, the volume in which the EMF may be assumed to be more or less homogeneous is very limited. The inhomogeneity may limit the accuracy of the standard much more than factors discussed in section 6.6. The inhomogeneity increases when an OUT is

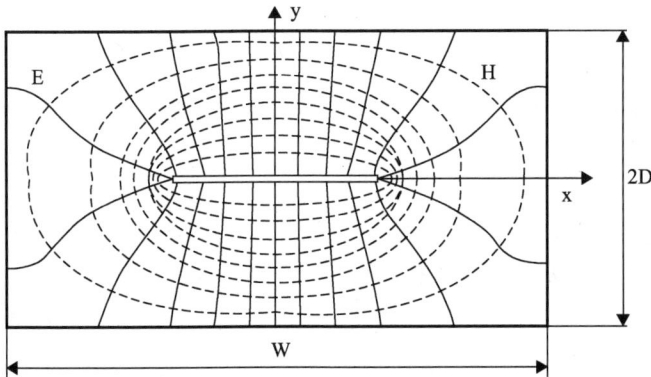

Figure 6.19   EMF distribution in a cross section of a screened symmetric strip line

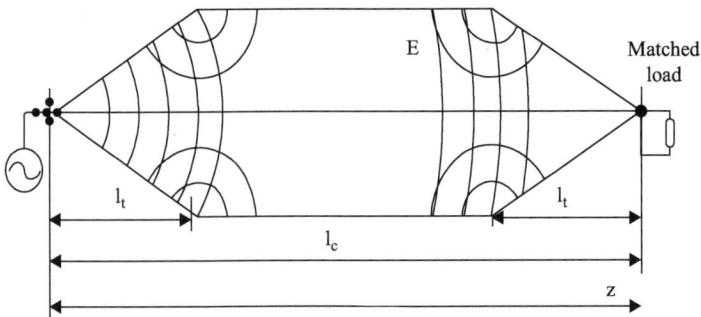

Figure 6.20   E-field distribution within and near matching transformers

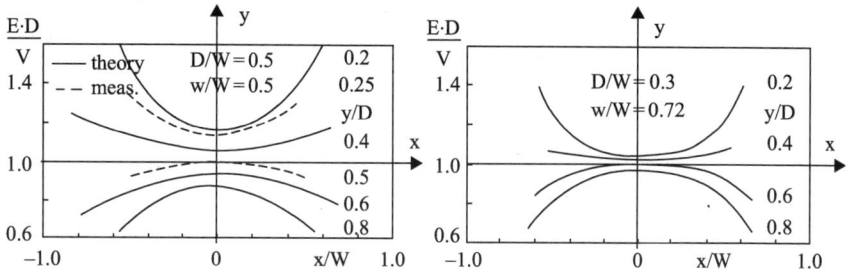

Figure 6.21 *E-field variation in a cross section of a symmetric strip line, referred to the field in the center of the line, for several y/D ratios*

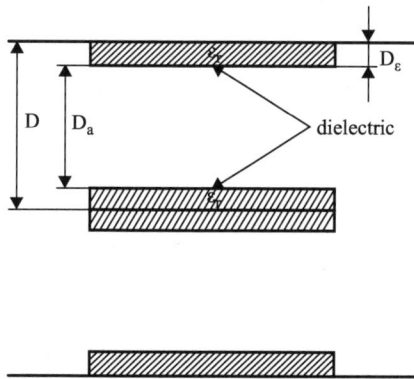

Figure 6.22 *A strip line with walls covered by a dielectric material*

placed in the line and, as mentioned, with increasing frequency and the resulting need to reduce the line dimensions. The EMF inhomogeneity, even when caused by quite a large OUT placed in the line, does not exclude an application of the line in the role of primary and/or secondary EMF standard. However, there is a need to assign theoretically or experimentally determined correction factors that would allow reference of the EMF inside the line to the plane wave conditions in free space. The procedures are troublesome, but they must be applied even if the line is applied for the exposure of small objects, if the accuracy of the exposure (EMF intensity) is of concern.

Calibration and exposure errors are of systematic and deterministic character. Thus, the use of correction factors may well improve the accuracy of procedures for which the line is used. There is a possibility, worked out by the authors, that allows an improvement in the EMF homogeneity in a line. The concept is based upon a placement of dielectric sheets at the plates of the line, as shown in Figure 6.22. The dielectric limits dispersed fields at the sides of the line and focuses it in the center of the line. In this case, the E-field in the center of the line is given by (6.38) [45].

$$E = V \frac{\varepsilon_r}{\varepsilon_r D_a + D_\varepsilon} \qquad (6.38)$$

where: ε_r – relative permittivity of the dielectric,
  D_ε – thickness of the dielectric layer,
  D_a – thickness of the air gap.
  Of course, $D_\varepsilon + D_a = D$.

Although the method presented for homogeneity improvement gives quite good results, unfortunately, it is unable to limit EMF deformations caused by an OUT immersed in the line. EMF disturbances caused by an OUT are, first of all, dependent on the relationship of the object's size to the line's dimensions. A limitation in the object's size is the most effective way to limit the effect caused by it upon the line. This effect is threefold in nature:

- EMF distribution disturbances in the line,
- local change of the line's wave impedance,
- alternation of the EMF intensity around the object.

The effects are caused by any object immersed in the line. In discussions devoted to the accuracy of the E- and H-field probe calibration in the line, this problem was ignored. This was due to an assumption that the field deformations caused by a probe in a line during probe calibration are of a similar order to the measured field with the use of the probe. Thus, the deformations should not cause remarkable limitations in measurement accuracy. However, this assumption is not fully correct, as the probe affects the line and, inversely, the line affects parameters of the probe (this is considered in the next chapter), and this leads to the necessity of checking the effects and the use of correction factors, as mentioned above. It is impossible to present a generalized analysis that would be valid in all calibration (and exposure) cases. Analysis should be performed for any specific case. To illustrate EMF deformations caused by a probe, an example is presented in Figure 6.23. The deformations in the case of exposure of larger objects are proportionally larger.

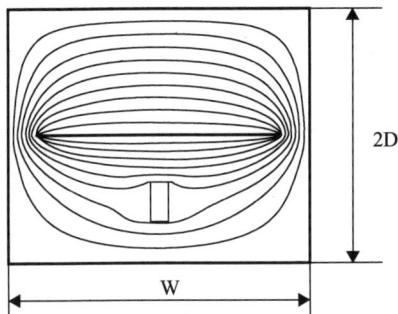

Figure 6.23   E-field deformations in a cross section of a line caused by an EMF probe

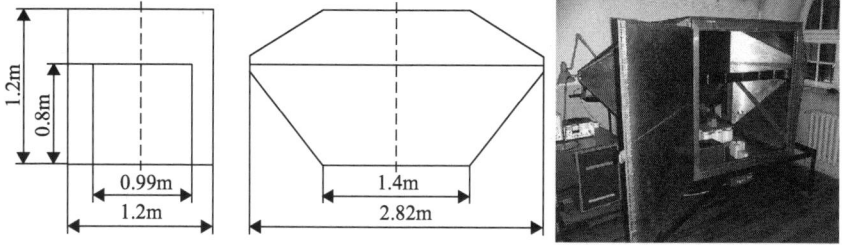

Figure 6.24   A screened unsymmetrical TEM cell with enlarged applicable volume

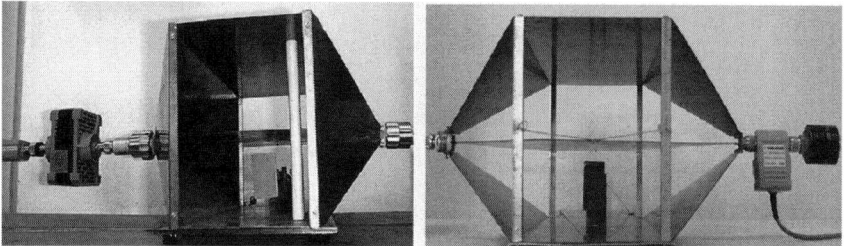

Figure 6.25   E-field probes during calibration in a TEM cell

In section 6.3, the problem of line asymmetry was discussed. As mentioned above, especially in the case of exposure systems, a problem appears with the usable volume in the line. An example of the volume increase by intentional line asymmetry is illustrated by the design presented in Figure 6.24. The figure shows sizes and a view of such a design applied in the authors' lab for purposes of investigations related to EMC.

6.8   An OUT in a TEM cell

So far, discussions presented have concentrated on effects caused by an OUT immersed in a line based on the line parameters. Now we will consider an inverse phenomenon, i.e., effects caused by a presence of material media, namely the walls of a line, upon parameters of the object.

6.8.1   An EMF probe in a TEM cell

An E-field probe immersed in a TEM cell during calibration is presented in Figure 6.25. A schematic diagram and an equivalent circuit of the simplest E-field probe are shown in Figure 6.26.

The transmittance T of the probe, in the medium frequency range, where the frequency response of the probe is frequency independent, is given by:

$$T = \frac{V_d}{e_A} = \frac{C_A}{C_A + C_d} \tag{6.39}$$

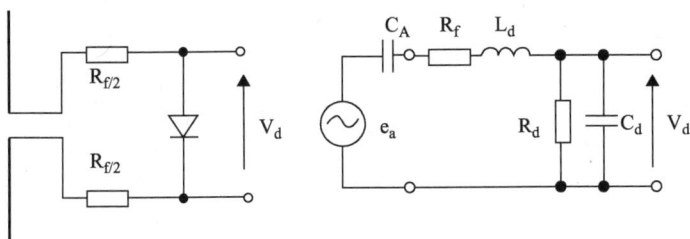

Figure 6.26   A schematic diagram and an equivalent circuit of an E-field probe

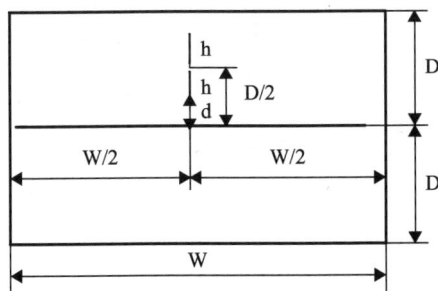

Figure 6.27   A dipole antenna in a TEM cell

where: e_A – electromotive force induced by the field in the antenna,
    V_d – voltage at the detector,
    C_A – input capacitance of the antenna,
    C_d – input capacitance of the detector.

In the conditions of calibration, for instance, in a fully screened, symmetric strip line, an antenna of the probe (usually a short symmetric dipole) is in close proximity to the line walls, as shown in Figure 6.27. The presence of the walls affects the input impedance of the antenna, or in this case its imaginary part.

An analysis of the effect caused by the conducting plates of the line upon the input impedance of the antenna requires taking into consideration the presence of an infinitely large number of images of the antenna in the line walls and mutual couplings among all the antennas. The mirror reflections are well known from antenna theory. However, in considering the antenna, only one reflection is taken into account: an antenna and its first image, e.g., in the earth. In the considerations for the first time the multiple reflections are and have to be taken into account. The system of reflections is shown in Figure 6.28 [21].

Because of the measured field integration, which is of essential importance when non-plane fields are measured (a situation typical in the case of near-field measurements), the size of the antenna should be as short as possible, which may be expressed in the form $h \ll \lambda$. We may assume here that the input impedance of the antenna is represented by the input reactance [basic condition while (6.39)

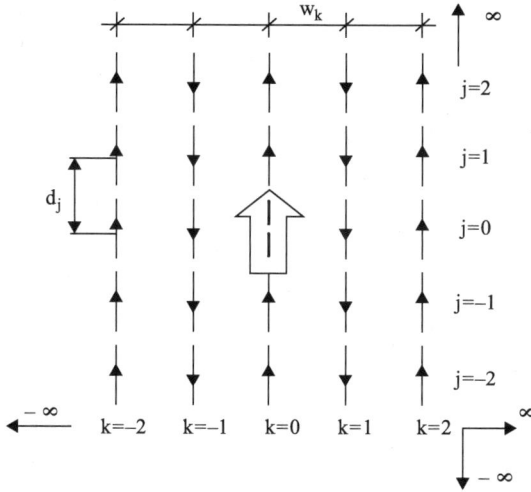

Figure 6.28   Mirror reflections distribution in a screened strip line

was introduced], or we may say it is a capacitive antenna. If we take into account the symmetry of the set of multiple reflections, the input reactance of the antenna X_A', presented in the screened strip line, is given by:

$$X_A' = X_A + \Delta X_A = X_A + 4\sum_{j=1}^{\infty}\sum_{k=1}^{\infty}X_{jk} \qquad (6.40)$$

where: X_A – input reactance of the antenna in free space,

ΔX_A – change of the input reactance due to the presence of mutual couplings of the antenna and its mirror reflections,

X_{jk} – mutual reactances of the antenna and its mirror reflections.

To illustrate the scale of the effect caused by walls of the line perpendicular and parallel to the antenna, appropriate errors $\delta_\perp$ and $\delta_=$ are introduced. These errors are defined as follows:

$$\delta_\perp = \frac{\Delta X_{A\perp}}{X_A} \quad \text{and} \quad \delta_= = \frac{\Delta X_{A=}}{X_A} \qquad (6.41)$$

where: $\Delta X_{A\perp}$ and $\Delta X_{A=}$ – mutual reactances in a row or column.

Equations (6.41) are plotted in Figures 6.29 and 6.30 for several slenderness ratios of the dipoles.

Calculated and measured input capacitances were obtained for a symmetric dipole antenna of length $2h = 0.06$ m, defined in way similar to calibration errors given by (6.41), i.e., for the antenna placed perpendicular to two conducting plates $\delta_{C\perp}$ and parallel to the plates $\delta_{C=}$. Results of calculations and measurements are presented in Figures 6.31 and 6.32. The measurements were performed using a method proposed by the authors. The method is based upon the measured antenna

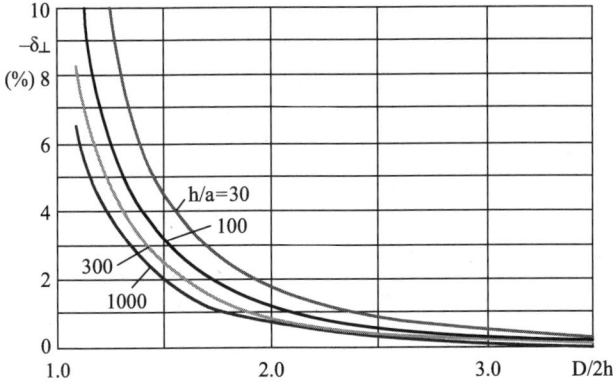

Figure 6.29 Error $\delta_\perp$ as a function of D/2h

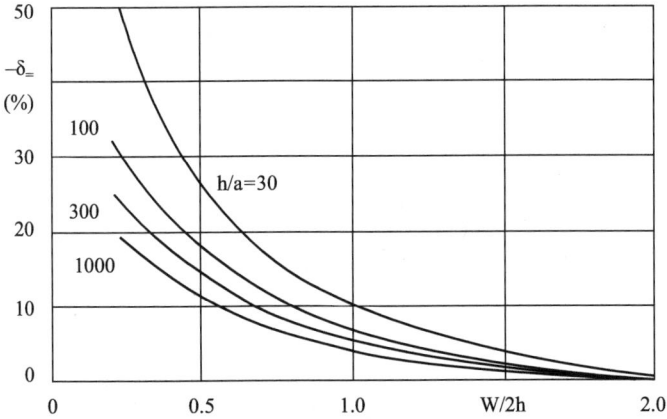

Figure 6.30 Error $\delta_=$ as a function of W/2h

as a part of a resonant circuit of a small generator connected to the input of the antenna. A schematic diagram of the generator is shown in Figure 6.33. The generator is closed in a metallic box and does not need any connection with outside devices. Its frequency changes are measured at a distance by a spectrum analyzer.

The analysis presented is of primary importance when an EMF probe is calibrated in a TEM cell or similar device. It allows calculation (measurement) of the correction factors necessary to correct effects caused by the presence of conducting plates and to have results of calibrations that would be equivalent to free space conditions.

As with almost all of the discussion presented here, the analysis and/or measurements have to be performed for a specific case of a standardization system and a calibrated probe. To illustrate this statement, we will return to (6.39). During a calibration, variations of C_A are of little concern, but the variations of transmittance

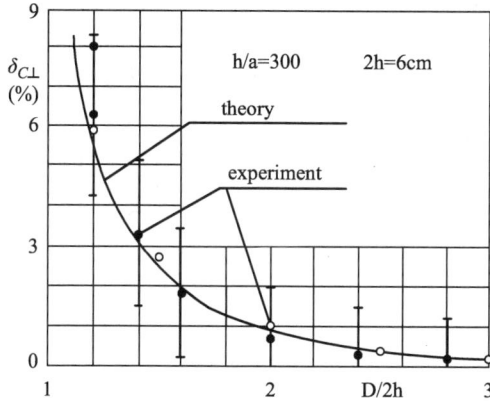

Figure 6.31 Alternations of the input capacitance of the antenna $\delta_{C\perp}$ versus D/2h

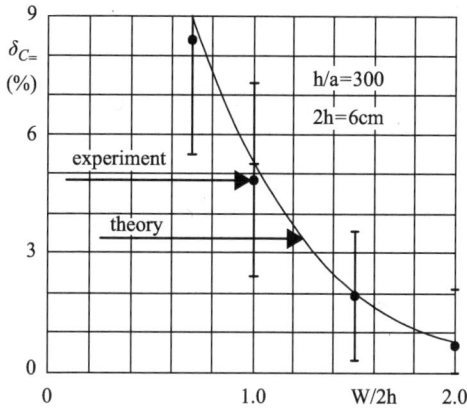

Figure 6.32 Alternations of the input capacitance of the antenna $\delta_{C=}$ versus W/2h

and, as a result, of the calibrated probe sensitivity, are of concern. Differentiating (6.39), we have:

$$\frac{\Delta T}{T} = \frac{\Delta C_A}{C_A} \cdot \frac{C_d}{C_d + C_A} \tag{6.42}$$

The first part of this equation reflects the presence of interactions of the calibrated antenna with a calibration system; the latter fraction limits the effect of the capacitive divider at the probe inputs (see Figure 6.26). It may be remarked here that the sensitivity of a probe increases with the increase of the probe's antenna input capacitance [see (6.39)], while its insensitivity to the presence of conducting media in its proximity increases with the increase in input capacitance of a detector. This shows once more the need for an individual approach to any individual case.

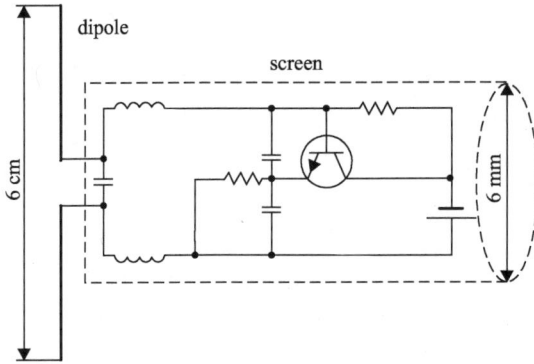

Figure 6.33   Schematic diagram of a generator applied for ΔC_A measurement

Every E-field sensor is sensitive to the presence of any material media in its neighborhood, regardless of the reason for the presence. This effect appears during a calibration as well as during measurement. A magnetic field sensor working at frequencies below its own resonance is quite insensitive to the effect. However, an H-field probe with a loop antenna working above its resonant frequency is sensitive. The transmittance T of the probe is given by:

$$T = \frac{A}{L_A} \tag{6.43}$$

where: A – constant,
    L_A – inductance of the loop.

It may be concluded from this equation that the probe is sensitive to the antenna's inductance variations, which may be of the same nature as discussed above in the case of E-field probes. In order to not repeat the discussion as presented above, only final results of estimations and measurements of relative inductance measurements are presented here. Figure 6.34 presents results of

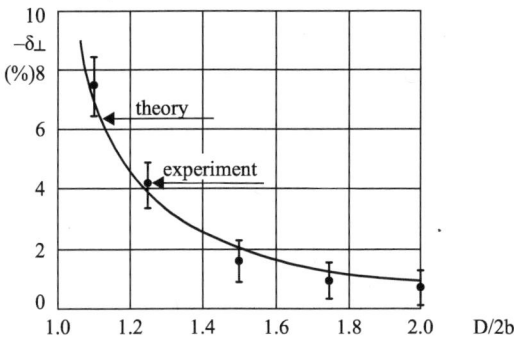

Figure 6.34   Calculated and measured values of $\delta_\perp$ versus D/2b

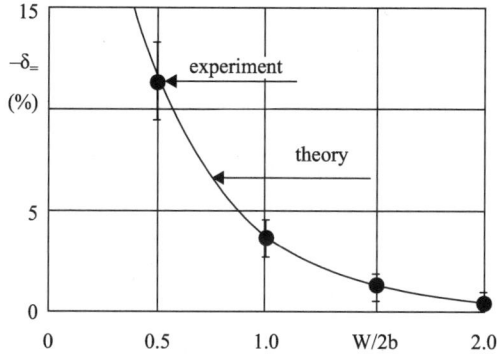

Figure 6.35   Calculated and measured values of $\delta_=$ versus W/2b

estimations and measurements of relative inductance variations ($\delta_\perp$) when a loop antenna is placed perpendicularly to two conducting plates and similarly, in Figure 6.35, when placed parallel to the plates ($\delta_=$).

Similar to the case of the E-field probes, the analyses and measurements of H-field probes calibrated in a line should be done separately for any particular type of line and the probe calibrated in it. We may add here that calibration of H-field probes using a TEM cell or similar device is done rarely. For this purpose, a standard with loop antennas is more convenient; however, the effect exists and in order to complete the presented considerations it has to be mentioned.

One more aspect of the problem should be mentioned. In the near field, EMF surveying uses probes of a spherical pattern. This pattern is a result of a spatial combination of three mutually perpendicular probes of sinusoidal pattern, and the solution makes it possible to simply measure the EMF of quasi-spheroidal polarization, i.e., EMF that has three spatial components. This concept applies to E-field and H-field probes alike. These probes are often calibrated in linear polarized EMF in a TEM cell because of the simplicity, sufficient accuracy, and standard approach to using such a procedure. However, the procedure requires taking into account that the interaction between the cell and any of the three components of the probe are different and are a function of the cell sizes and spatial placement of three antennas in the probe. Checking the shape of the pattern in the given conditions may lead to false conclusions. Such checking may be done in linear polarized EMF. However, it is allowable when the sizes of the checked probes are small in relation to the distance between walls of a line used for this purpose. The best solution here would be a spherically polarized standard EMF. A block diagram of a spherically polarized H-field set, designed and proven by the authors, is shown in Figure 6.36. The idea of the concept is based upon the use of three perpendicular loops fed from an HF source in such a way that currents in separate loops are as follows:

$$I_1 = A \cos \Omega t$$
$$I_2 = B \cos \Omega t \sin \omega t \qquad (6.44)$$
$$I_3 = C \cos \Omega t \cos \omega t$$

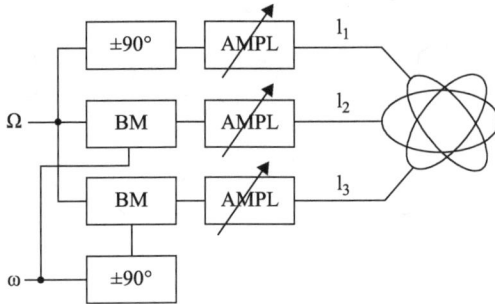

Figure 6.36   Block diagram of quasi-spheroidal polarized H-field source

where: A, B, and C – amplitudes,
Ω – carrier wave frequency,
ω – modulating frequency.

The quasi-spheroidal or quasi-ellipsoidal polarization is obtained as a circular or elliptical polarization rotating in space with frequency ω. A similar concept was proven as an E-field source. Limitations in its use and estimated accuracy show that it is rather a presentation of technical possibility than a solution of practical importance. Thus, the TEM cell or any similar device remains irreplaceable in the case of spherical probe calibration. Regardless, the conclusion may be shown that such an approach is irreplaceable in the role of an exposure system, especially for biomedical investigations on living objects. Energy absorption in the body is a function of mutual placement of E-field vector and the body. In the case of living objects, it is possible to limit their natural behavior; however, this could cause bioeffects exceeding that of the exposure. In order to not restrict the movement of investigated animals, and still ensure uniform EM energy absorption, a quasi-spherical EMF is suggested. At lower frequencies, this may be done in a set-up as shown in Figure 6.36 or a modified version with three pairs of plate electrodes; at higher frequencies a system with three log-periodic antennas (Figure 6.37) ensures better EMF uniformity and works in a wide frequency range. Good field uniformity in this case may be assured if distances between antennas and the OUT are $R > 2D^2/\lambda$, which represents far-field conditions.

In any case of such quasi-spherical EMF polarization, it is possible to have an EMF that is linear, circular, elliptical, quasi-spherical, and quasi-elliptical polarized, of arbitrary spatial position, by regulation of amplification and by introduction of phase regulation by phase shifters instead of a constant $\pm 90°$ shift.

6.8.2   A TEM cell as an exposure system

Considerations related to the effects caused by an EMF standard upon an antenna may be summarized as follows:

- The presence of conducting plates of a line (TEM cell) affects the input impedance of a calibrated antenna.

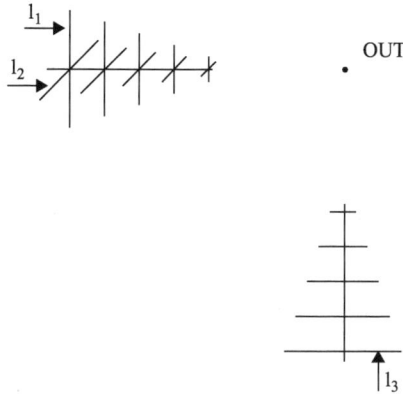

Figure 6.37   Quasi-spheroidal EMF generation at high frequencies

- As a result, the sensitivity of the probe is affected.
- Due to the deterministic character of the phenomenon, there is a possibility of introducing appropriate correction factors that would allow improvement of the accuracy of calibration and to refer it to free space conditions.
- The relatively small size of EMF probes allows any effect on the cell to be ignored in many cases.

The presence of the phenomenon in the case of an antenna suggests that a similar interaction will appear when any other material object is investigated in a line. Consider the role of the phenomenon in the case when the line is used for larger material objects or more than a single object. Especially in biomedical experiments, as many objects as possible are usually placed in an exposure system. Unlike antenna calibration, where it is possible to assume that the antenna is a single size ($h \gg a$) and perfectly conducting, a biological object exposed in a TEM cell is a three-dimensional object, usually the dimensions are irregular, and the electrical structure of the object is inhomogeneous. In terms of its electrical properties, the object may be described as semiconductor or lossy dielectric. Unlike with antennas, an "input impedance" probably exists; however, it is not well defined. Thus, analysis of the interaction of the object and its mirror reflections should be based upon different magnitudes. Moreover, again in comparison to an antenna, there is no simple approach to using theoretical methods. Fortunately, available numerical codes allow estimation of the scale of the phenomenon.

There are many numerical models of living beings, from the simplest to "millimeter resolution" objects [23]. They are widely applied in bioelectromagnetics. Unfortunately, up to now no one has tried to consider the scale and importance of the interactions discussed. To illustrate the scale, several examples are presented below in which the simplest model is in use [13]. In the first case, a cube of sides $h = 1.5$ cm, conductivity $\sigma = 1$ S/m, and $\varepsilon_r = 80$ was used. Its electrical parameters represent typical values for living matter. As it was impossible to use the input impedance, the power absorbed by the object was estimated. The object was immersed between two plates and, disregarding the distance between them, the electric field was kept constant at $E = 1$ V/m. Figure 6.38 presents results of estimations when a single object is exposed in the cell. The estimations were performed using two different codes that additionally permit comparison of the concordance of obtained results. Although there is a disagreement between results when the distance between the line plates approaches the size of the cube, it may be seen from the figure that power absorbed by the object approaches the value estimated for a free-space exposition when $D > 2h$.

The second example shows three mice in a line. In this case, each mouse is represented by a cube of side $h = 5$ cm, $\sigma = 0.84$ S/m, and $\varepsilon_r = 80$. As previously, the estimations assumed constant intensity of E-field at a level of 1 V/m. The mice are placed between the cell walls as shown in Figure 6.39. Results of calculations are presented in Figure 6.40.

A traditional approach to absorption, in the case of animals' exposure, is based upon measurement of the power absorbed in the cell, and individual absorption is calculated by dividing the power by the number of exposed objects. This approach assumes that power absorbed by each object is identical. An evident weakness of the assumption may be seen from the above presented results of estimations. The estimations were performed for very simple conditions. However, they clearly show the importance of a problem never considered before. Similar estimations performed for conditions when there are as many animals as possible in the cell

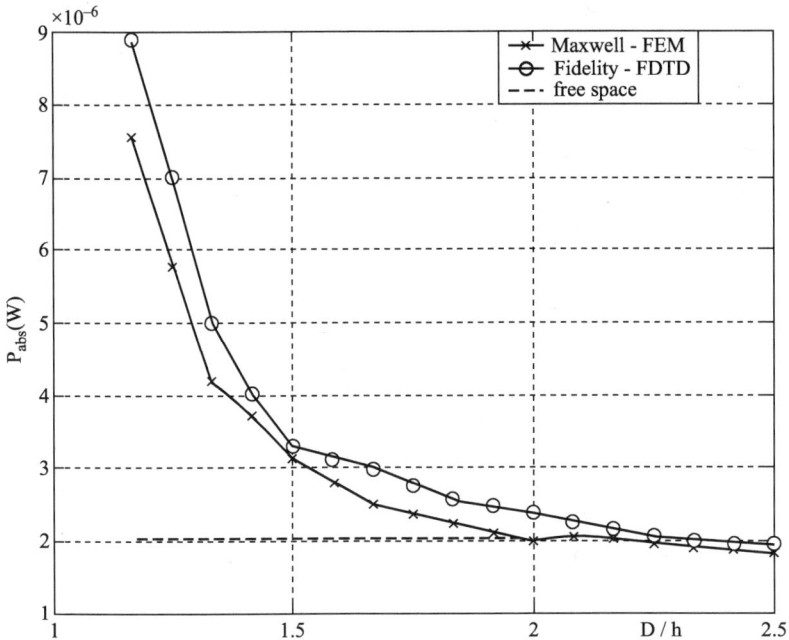

Figure 6.38 · Power absorbed by a single cube versus D/h

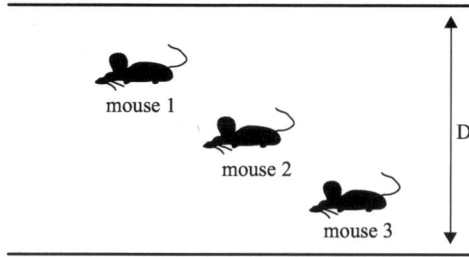

Figure 6.39   Three mice in a cell

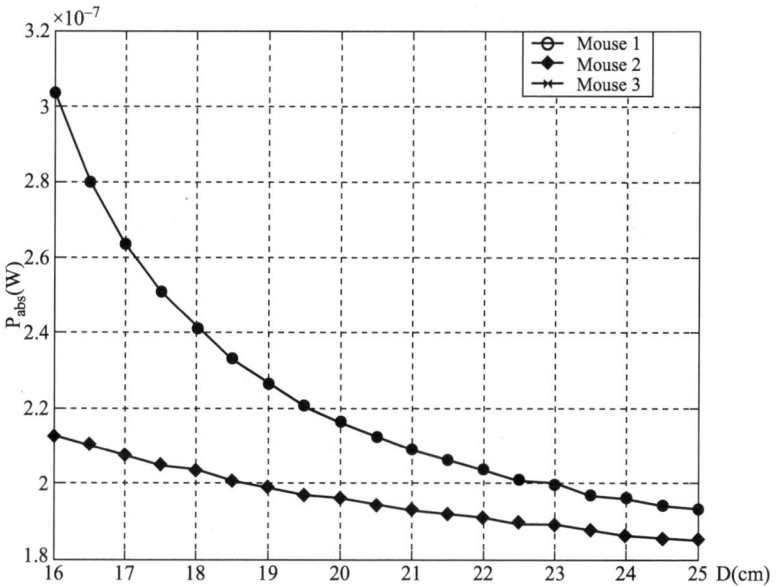

Figure 6.40   Power absorption in mice versus distance between cell walls

show that the difference in power absorbed by the animals may exceed an order of magnitude or more. Moreover, even in the simplest conditions, when a single, homogeneous object is exposed, distribution of the absorbed power is nonuniform (see Figure 6.41). This leads to the conclusion that the majority of biomedical investigations are of a qualitative character rather than quantitative. This may be one reason why data in the literature shows remarkable differences in the results of experiments performed under "identical conditions." The differences may be larger when the results of exposure upon a selected organ or tissue are being investigated.

Of course, a TEM cell and similar solutions, because of their advantages, were, are, and will be applied in a variety experiments as exposure systems. This is fully understandable and acceptable. However, the cell, like any other tool, has limitations in its use. It is to aid in understanding the limitations and their reasons that the

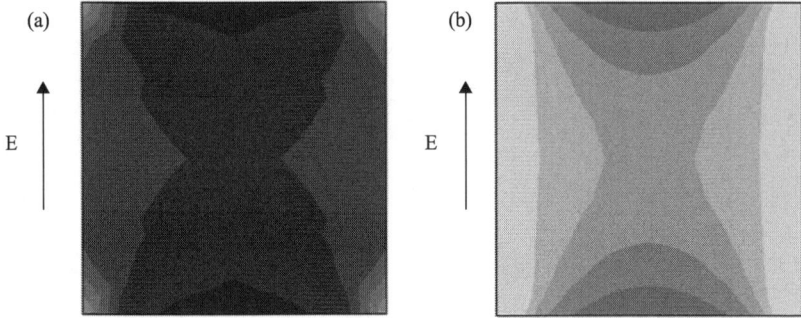

Figure 6.41   Results of estimations of the E-field distribution: (a) f = 100 MHz, (b) f = 1 GHz

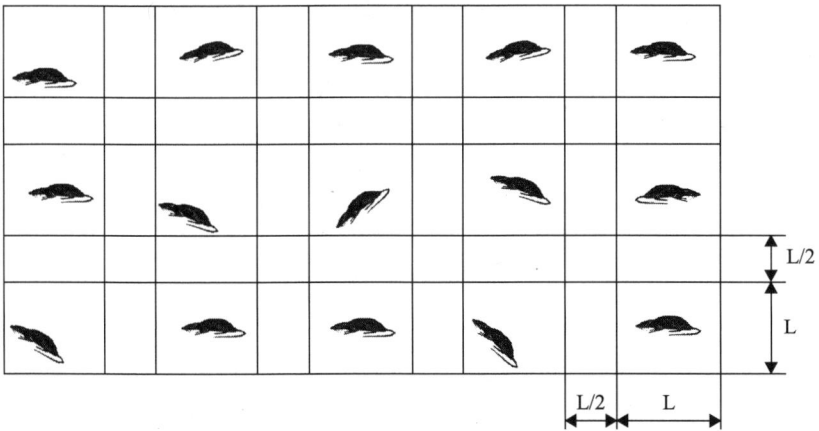

Figure 6.42   Proposed cage for animal exposure in a TEM cell

above considerations are presented. For instance, exposure of animals, while mutual couplings are negligible, may be performed only when the animals are in a single layer, in the center between the cell walls, and the animals are separated from each other by a distance around of half its size. Such a solution was proposed by the authors, and a sketch of a proposed "cage" is shown in Figure 6.42. Square cells of sizes L × L are designed for investigation of animals. All the cells are separated from each other by "empty" corridors of width L/2. According to our estimations, this should ensure an acceptable level of accuracy. The cage may be immersed in a TEM cell, parallel to its walls and in the center between them. However, the maximal size of the cage should not exceed 0.5w in width in order to achieve uniform exposure of the objects (see Figure 6.21). In the case of a symmetric TEM cell, as shown in Figure 6.20, there is a possibility to immerse two cages in the cell, one between the center conductor and the upper sidewall and the other below the side conductor. In the case of exposure of the objects, for instance, by a horn antenna, the radiation should be perpendicular to the cage. The cage is made of low-loss dielectric material that allows losses to be ignored. The bottom of the cage is made of a plate of the same material and, if necessary, a similar plate is applied to cover the cage.

One comment more: This analysis takes into account only one phenomenon, i.e., the interactions between an exposed object and an exposure system. When complex accuracy estimations are performed, it is necessary to take into consideration any other factor limiting the accuracy of exposure, specific for the type of exposure system and exposed objects.

6.8.3   Absorption and polarization

Considerations presented in section 6.8.2 revealed the role played by mutual coupling between OUTs and couplings between OUTs and the exposure system. We remind the reader that these couplings can lead to large errors, especially when the exposure system is maximally loaded with the objects. Regardless, it is necessary to mention that EMF polarization in the cell is linear. This may lead to another error in EM energy absorption estimations resulting from field polarization. As is evident, and as already mentioned in section 3.5.3.2, absorption depends on the mutual positioning of an object and the field components. This issue may be disregarded when nonliving objects are of concern. When living objects are exposed, there are two possibilities:

- Limitation of the possibility of movement by the animal. The solution may ensure quite uniform and estimable absorption, but at the expense of additional effects caused by the restriction, which may dominate in the experiment.
- Allowing free behavior of the animals, which would require a more complex (and expensive) exposure system.

To show the order of possible errors here, and to caution against them and show the way to limit them, we bring several estimations of absorbed energy in bodies exposed to different polarizations. Let's imagine a cage, as in Figure 6.42, containing nine cells in three rows. The cage is illuminated by a half-wave dipole antenna as in Figure 6.43. In every cell there are objects of size $L/3 \times L/3 \times L$,

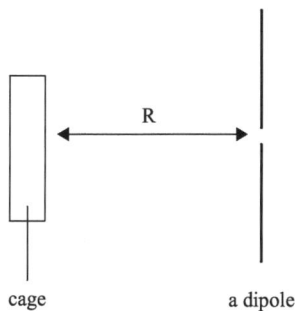

Figure 6.43   A cage illuminated by a half-wave dipole

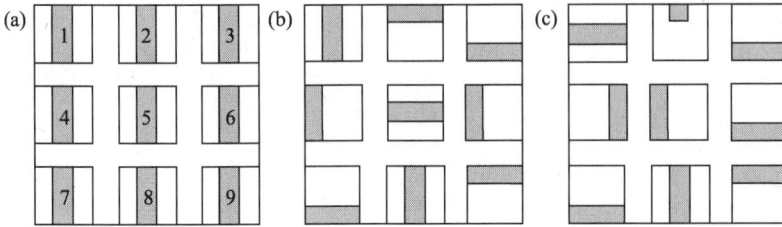

Figure 6.44   Objects positioning in the cage, as described in the text

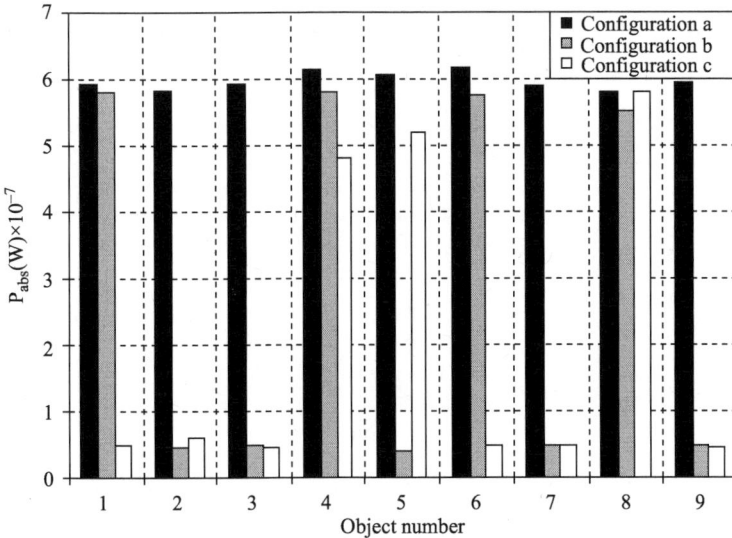

Figure 6.45   Absorbed energy in configurations shown in Figure 6.44

placed symmetrically in the center of the cell, with similar parameters as assumed in section 6.8.3, frequency 100 MHz, and assumed E-field intensity close to the cage of 1 V/m. Distance R between the cage and the dipole should ensure quite uniform EMF distribution at the plane of the cage. Three versions of the objects' position in the cage were considered. They are shown in Figure 6.44. In Figure 6.44a, each object is positioned parallel to the E-field vector. In Figure 6.44b, only objects 1, 4, 6, and 8 are parallel to the vector; and in case c, objects 4, 5, and 8. Estimated energy absorption in the objects is shown in Figure 6.45.

As may be deduced from the figure in configuration a, the absorbed energy is almost identical in each object. In the cases where the main axis of an object is perpendicular to the E-field vector, the absorbed energy is less by almost an order of magnitude.

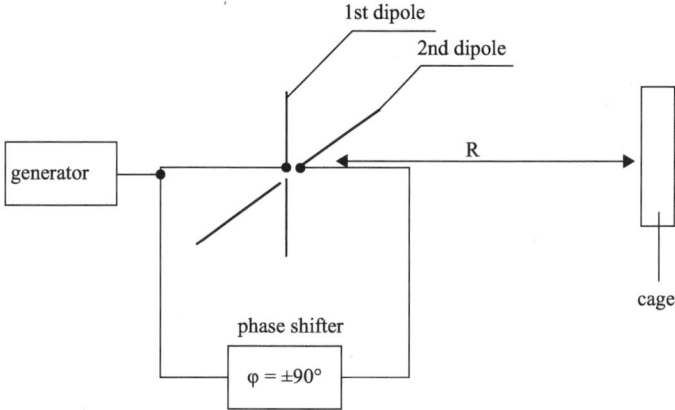

Figure 6.46   Circular polarized EMF generation

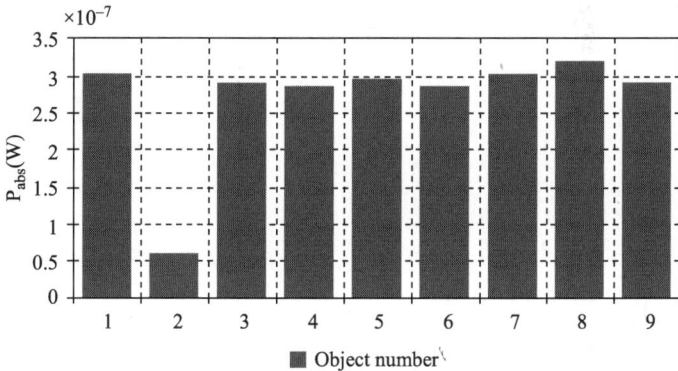

Figure 6.47   Absorbed energy when the cage is illuminated with circular polarized EMF

Now let us consider the same case of objects distributed in the cage as shown in Figure 6.45c. Now the cage is illuminated by a circular polarized EMF using, for instance, the set-up shown in Figure 6.46. Results of estimates are shown in Figure 6.47.

Finally, the cage in Figure 6.44c is exposed to a spherical polarized EMF. Results of estimations are shown in Figure 6.48. This figure does not need any comment. It clearly shows that, in order to have uniform energy absorption, without regard to the spatial position of an OUT, it needs to be exposed to a spherical polarized EMF.

It is necessary to explain here that spherical polarization is a physical fiction and does not actually exist! Regardless, in estimations, as with the results presented in Figure 6.48, such a polarization was assumed. However, this does not mean that

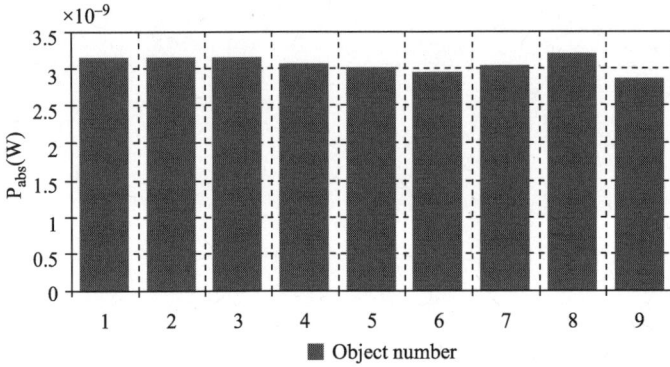

Figure 6.48   Absorbed energy when the cage is illuminated with spherical polarized EMF

the conclusion on uniform absorption is invalid. It means only that, in order to have uniform absorption, regardless of an OUT's positioning in relation to spatial EMF components, in set-ups such as those proposed in Figures 6.36 and 6.37, it is necessary to reduce I_1, in relation to two other currents, by $\sqrt{2}$ [see (6.44)].

Chapter 7
Accuracy analysis of the standards with horn antennas

Section 4.6 presented an analysis devoted to the accuracy of directional antenna calibration based on a log-periodic antenna. The method presented there may be fully applied to calibration of other types of directional antennas, including horn antennas, slot antennas, and others. In EMF metrology, horn antennas are widely used. However, because of their specific nature, the method is of limited applicability.

1. Log-periodic and horn antennas work in different frequency ranges. The former are used at frequencies below 1 GHz, while the latter above that range. Above 1 GHz the construction of EMF standards using standard dipole antennas is troublesome because of quite short waves and resulting problems of a technical nature in the construction of standard antennas of required slenderness. Moreover, use of a standard dipole antenna as a directional one is inconvenient because of its very low power gain. This validates the application of horn antennas as standard receiving and transmitting antennas at frequencies above 1 GHz. While the frequency limit in the case of log-periodic antennas results from problems with the construction of small dipoles, in the case of horn antennas the limit results from the large sizes of the antennas below 1 GHz. For example, the size of horn antenna type 3160-01 for frequency range 0.96–1.45 GHz, manufactured by EMCO, is comparable with that of a log-periodic antenna for frequency range 200–1000 MHz, around 1 m. Reducing the lower corner frequency of a horn antenna in order to reduce its size is usually achieved by a special design of the antenna. An example of wideband horn antennas of reduced sizes, working within the frequency range 1–18 GHz and 18–40 GHz, is shown in Figure 7.1. To illustrate the sizes of the antennas, a typical 9V battery was placed between them.

2. Log-periodic antennas are, as a rule, fed from coaxial cables, whereas horn antennas are usually fed from a waveguide; the simplest "horn antenna" is an open end of a waveguide. The horn plays the role of a transformer, matching the wave impedance of a waveguide to the intrinsic impedance of free space, illustrating the similarity of technology in antennas and waveguides. The specificity of the technology affects the wideband nature of horn antennas, which usually do not exceed a single octave, whereas log-periodic antennas

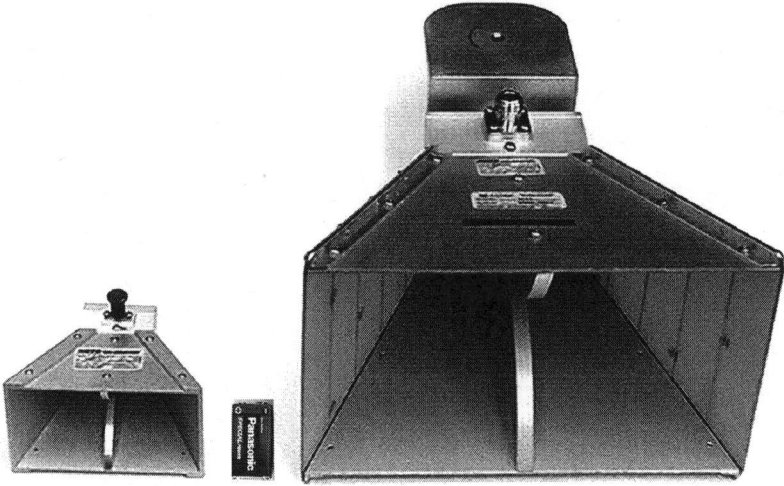

Figure 7.1   Wideband horn antennas manufactured by EMCO

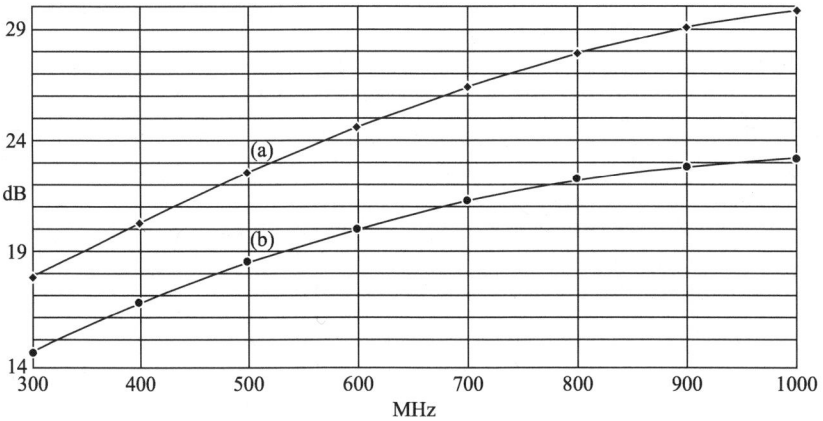

Figure 7.2   Measured antenna factor of the DAMZ-4/50 type log-periodic antenna as a function of frequency: (a) with 12 m long cable; (b) without the cable

may exceed one decade. Moreover, the gain of horn antennas is much larger compared to the log-periodic antennas. This statement is confirmed by a comparison of data presented in Figure 7.2 and in Tables 7.1a and 7.1b.

3. The calibration of a log-periodic antenna consists of measurement of the antenna factor AF (this term is traditionally tied to the receiving antennas and indicates the ratio of E-field at the location of the antenna to the output voltage V across the load connected to the antenna $AF = E/V$). Results of the

Table 7.1a Parameters of EMCO horn antennas applied by the authors

EMCO antenna type	Frequency range (GHz)	Antenna factor AF (dB)	Gain (dB)	VSWR
3160-01	' 0.96–1.45	15.3–15.6	28,8–60,8	1.6
3160-02	1.12–1.70	16.7–17.0	28,0–59,9	1.5
3160-03	1.70–2.60	20.4–20.7	27,4–59,6	1.3
3160-04	2.60–3.95	23.7–23.8	30,5–67,4	1.3
3160-05	3.95–5.85	27.2–27.4	31,08–65,43	1.3
3160-06	5.85–8.20	29.8–30.0	37,1–70,2	1.3
3160-07	8.20–12.40	33.4–33.6	32,3–70,2	1.3
3160-08	12.40–18.00	37.0–37.3	31,9–63,5	1.4
3160-09	18.00–26.50	40.2–40.5	32,4–65,3	1.4
3160-10	26.50–40.00	43.4–43.7	33,7–70,39	1.4
3115	1–18	26–47	5.5–16	1.5
3116	18–40	45–50	11–16	1.6

Table 7.1b Parameters of Flann Microwave horn antennas applied by the authors

Flann Microwave model no.	Frequency range (GHz)	Gain (dB) accuracy ±0.25 dB
20240–25	17.6–26.7	23.9–27.1
22240–25	26.4–40.1	23.9–27.2
24240–25	39.3–59.7	23.1–26.5
25240–25	49.9–75.8	23.9–27.1
27240–25	73.8–112.0	23.9–27.2

measurements of a DAMZ-4/50 type antenna, performed using methods presented in Chapter 4, are shown in Figure 7.2. The effective area and gain of a horn antenna are of concern. Contrary to the antenna factor, both magnitudes may be precisely calculated analytically.

4. Within the frequency range where log-periodic antennas are in use, the most often applied magnitude is the electric field strength E, while at microwave frequencies, power density S dominates. This reflects the units used in further considerations.

7.1 Accuracy of determination of the power gain

The accuracy of the power gain and the effective area of the horn antenna are of primary importance for the class of the EMF standard in which the antenna is applied. Possible estimation errors and measurements are discussed in this chapter.

Because of geometrical sizes, at the lowest frequencies, open-ended waveguides are often used as radiators. Such radiators, such as that in the National Institute of Standards and Technology in Boulder, are used at frequencies between 200 and 500 MHz. Although such a solution is quite large, its main advantage is the ability to compare EMF standards with dipole antennas and with directional antennas over a much wider frequency range than only at a "border frequency" of 1 GHz. Apart from the size of the standard, its main disadvantage is a quite large reflection factor at the radiator input due to differences between the wave impedance of the waveguide for the mode applied (Z_w) and the intrinsic impedance of free space (Z). Taking this into consideration, it is possible to express the reflection factor Γ in the form:

$$\Gamma = \frac{Z - Z_w}{Z + Z_w} \tag{7.1}$$

For basic mode H_{10}, the wave impedance Z_{w10} is:

$$Z_{w10} = Z\frac{\lambda_w}{\lambda} = \frac{Z}{\sqrt{1 - (\lambda/2a)^2}} \tag{7.2}$$

where: λ_w – wavelength in the waveguide,
a – size of the wider wall (in the case of a rectangular waveguide).

Apart from the problems with radiator (waveguide) matching, which should be done at each separate frequency, there are problems with the measurement of excitation. Thus, such a device is used only for verification of transfer standards and comparative measurements at several selected frequencies. In regular procedures, the transfer standards are applied rather than the open-ended waveguide.

The power gain, G, of a radiator in the form of an open-ended waveguide is given by [18]:

$$G = 21.6f \cdot a \tag{7.3}$$

or

$$G_{[dB]} = 10 \log(f \cdot a) + 13,34 \tag{7.4}$$

where: f – frequency (GHz).

The accuracy of the power gain estimation using the above equations is within ± 0.5 dB, which is acceptable for distances d (in m) from the aperture larger than 2a [m]. The agreement of calculated and measured values of the power gain is compared in Figure 7.3 for a frequency of 500 MHz, as a function of distance d for a = 53.34 cm [8].

In standards for the frequency range 1–40 GHz, used by the authors, EMCO horn antennas for different frequency subranges, are set out in Table 7.1a; the antennas are shown in Figure 7.4. The wideband horns are used for ranges 1–18 GHz and 18–40 GHz for subsidiary measurements and as transmitting antennas. A similar solution is suggested by the Polish standard. The accuracy of the power gain estimations, required by the standard, is within the range of $\pm 0.3 - \pm 0.5$ dB.

Similar to the case of TEM cells, where matching was achieved using matching transformers, the horn plays the role of the matching system that allows

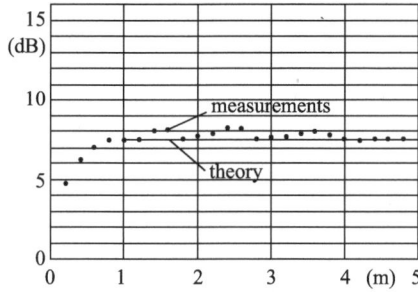

Figure 7.3   A power gain of an open-ended waveguide radiator at 500 MHz versus d

Figure 7.4   A set of EMCO horn antennas for frequency range 0.96–40 (left) and Flann Microwave antennas for frequency range 18–112 GHz (right) are shown; in both photos a standard 9V battery is added in order to visualize dimensions of the antennas

accurate matching of the wave impedance of a waveguide to that of the free space and then a matching of the waveguide to other components of a microwave track (bandpass filters, reflectometers, power sources, matched loads, etc.) in the case of horn antennas. The values of the VSWR of antennas applied by the authors, provided by the manufacturer, are indicated in Table 7.1a.

Within the frequency range 18–112 GHz, the authors use antennas manufactured by Flynn Microwave. Data on the antennas, supplied by the manufacturer, are given in Table 7.1b and the antennas are shown in Figure 7.4.

The power gain of an antenna is defined in the form:

$$G = \eta D \tag{7.5}$$

where: D – directivity of the antenna

     η – radiation efficiency, $\eta = \frac{P_r}{P_e}$

     P_r – power radiated,

     P_e – power exciting the antenna.

The power exciting the antenna is the difference between the power fed to the antenna and that reflected from it. The power balance at the antenna may be measured with acceptable accuracy, as discussed in the next section. In a first approximation it is possible to assume that:

$$G \approx D \tag{7.6}$$

and then estimate the antennas directivity using

$$D = \frac{4\pi}{\lambda^2} \eta_{ap} A_p \tag{7.7}$$

where: A_p – physical aperture of the antenna (as shown in Figure 7.5)
$\quad\quad \eta_{ap}$ – a coefficient of the aperture efficiency:

$$\eta_{ap} = \frac{E_{AV}^2}{(E^2)_{AV}} \tag{7.8}$$

where: E_{AV} – electric field averaged across the aperture (see Figure 7.5a)

Precise determination of the directivity and the power gain of a horn antenna is difficult, as the E-field distribution at the aperture of the antenna is not precisely known. Moreover, (7.7) is valid in far-field conditions. In the near field, the directivity and power gain are a function of the distance between the antenna and a point of observation. This leads to the necessity of determining the power gain with an approximation, which necessitates an estimate being made of the accuracy of the approximation and other simplifications applied. These estimations may then be proven experimentally.

For near-field conditions, when the distance d between the horn and a point of observation is less than that given by , more accurate formulas than (7.3) and (7.4) are used to estimate the power gain. The geometry of the antenna is shown in Figure 7.5 [32]:

$$G = 113{,}3 \cdot a \cdot b \cdot f^2 \cdot 10^{-R_H - R_E} \tag{7.9}$$

or

$$G_{[dB]} = 10 \log a \cdot b + 20 \log f + 20{,}54 - R_H - R_E \tag{7.10}$$

where: R_H – coefficient of the antenna's power gain reduction for plane H, a function of the antenna sizes, distance to the point of observation, and frequency:

$$R_H = 0{,}01\alpha(1 + 10{,}19\alpha + 0{,}51\alpha^2 - 0{,}097\alpha^3) \tag{7.11}$$

R_E – coefficient of the antenna's power gain reduction in plane E:

$$R_E = 0{,}1\beta^2(2{,}31 + 0{,}053\beta) \tag{7.12}$$

and

$$\alpha = \left(\frac{a^2 f}{0{,}3}\right)\left(\frac{1}{L_H} + \frac{1}{d}\right) \tag{7.13}$$

(a)

(b)

(c)

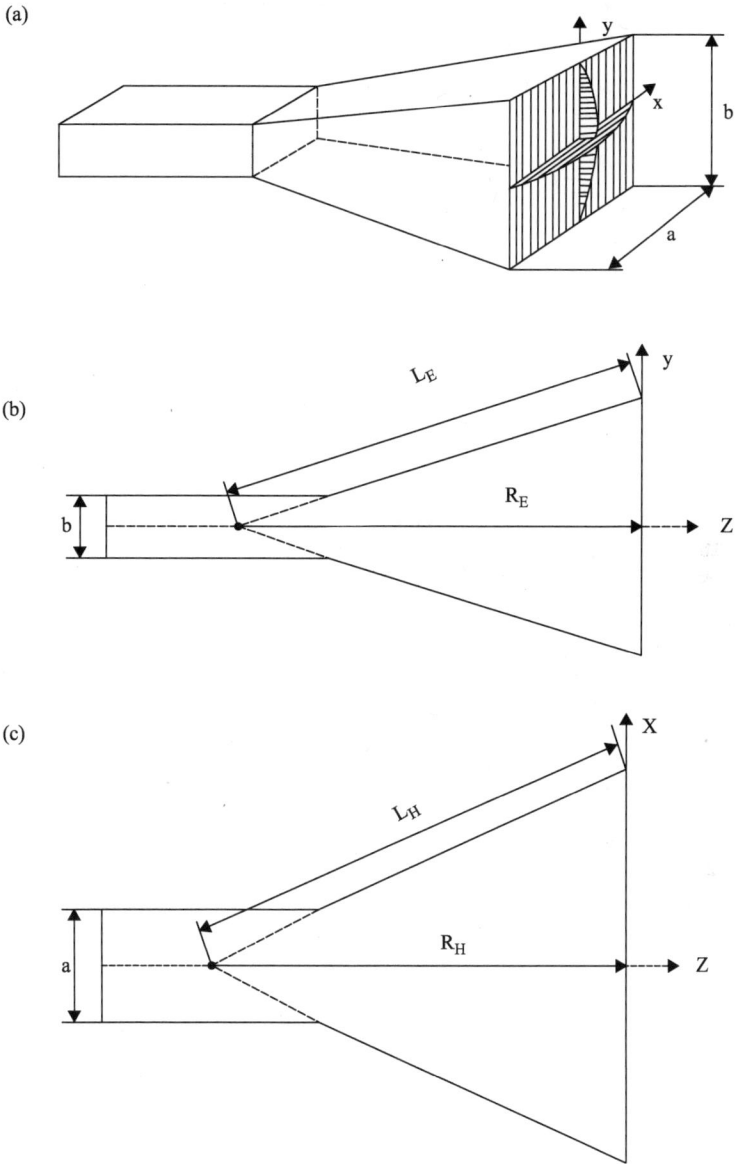

Figure 7.5 E-field distribution in the aperture of a pyramidal horn antenna indicated at the aperture of the antenna (a), dimensions of the horn in plane E (b), and in plane H (c)

or

$$\beta = \left(\frac{b^2 f}{0,3}\right)\left(\frac{1}{L_E} + \frac{1}{d}\right) \tag{7.14}$$

where: f – frequency [GHz]
      d – distance to a point of observation,
      L_E and L_H – slant radii,
      other indications as in Figure 7.5.

Inaccuracy in power gain determination in near-field conditions is estimated by different authors at the level of 0.5–1 dB; the maximal inaccuracy of the power gain estimations in standard horn antennas should not exceed ±0.1 dB.

The approaches presented above to standard horn antenna power gain estimations, determined theoretically, may be verified experimentally. The verification allows a precise determination of the gain of an antenna in any specific case of use, although taking into account all factors decreasing the accuracy of the gain estimations is impossible. For example, the precision of antenna manufacture, quality of the antenna layout, corrosion, aging effects, small mechanical defects, etc. are neglected in analytical methods of gain estimation, though their presence affects the results of measurement.

The most popular method of power gain measurement involves three different antennas or two identical ones [33]. The method is based upon a comparison of the power delivered to a load directly from a source and that through radiation by a transmitting antenna and received by a receiving antenna, as shown in block diagram in Figure 7.6. An appropriate choice of connecting cable length allows exclusion in the measurements of any errors caused by attenuation of the cables.

A product of the power gains of two directional antennas, polarizationally matched and coaxially placed on the axis of maxima of their radiation patterns, is given by:

$$G_t G_r = \left(\frac{P_r}{P_d}\right)\left(\frac{4\pi d}{\lambda}\right)^2 \frac{|1 - \Gamma_{ra}\Gamma_l|^2 |1 - \Gamma_{ta}\Gamma_g|^2}{(1 - |\Gamma_{ra}|^2)(1 - |\Gamma_{ta}|^2)|1 - \Gamma_g\Gamma_l|^2} \tag{7.15}$$

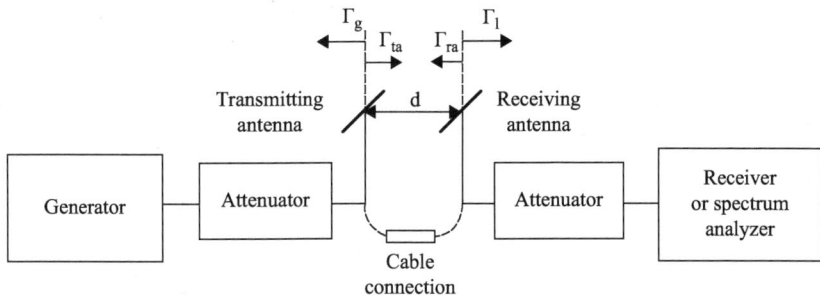

Figure 7.6   Block diagram of a set-up for power gain measurement

where: G_t – gain of the transmitting antenna,

G_r – gain of the receiving antenna,

P_r – power fed to the load from the receiving antenna,

P_d – power fed to the load through direct connection,

Γ_{ta} – reflection factor at the input of the transmitting antenna,

Γ_{ra} – reflection factor at the input of the receiving antenna,

Γ_g – reflection factor of the feeding track,

Γ_l – reflection factor of the receiving track.

The relationship of power fed to the load by the cables and through the "air" indicates the power losses in the space between the two antennas. During measurements carried out at a single frequency, it is possible to match the system in such a way that the reflection factors are as small as possible in order to be able to neglect them. When excitation of the transmitting antenna and the power delivered to the load are measured using directional coupler, the matching in the system is of secondary importance and may be omitted. In (7.15), there appear two unknown quantities, G_t and G_r. Thus, the equation may be used directly when two identical antennas are being measured. In the case of three different antennas, we will have a set of equations with three unknown quantities, i.e., the gains of the three antennas.

In the case of measurements using two identical antennas and checking that the matching in the track is sufficient and losses in the interconnections can be omitted, (7.15) takes the form:

$$G = \left(\frac{V_r}{V_d}\right)\left(\frac{4\pi d}{\lambda}\right) \tag{7.16}$$

where: V_r and V_d – voltages on the load of radiated and direct signals, respectively.

Works initiated in the area of near-field antennas measurement in early 1970s by Paul F. Wacker have been developed and currently these methods are widely used, especially for the measurement of parameters of large antennas [46]. Research in the area is presented in many papers. The above short summary is devoted only to checking of the power gain of directional antennas. In this instance, radiation pattern, side lobes, front to back ratio, and other parameters are not of concern if their presence does not affect calibration procedures.

7.2   Accuracy of excitation measurement

The accuracy of excitation measurement is of primary importance both during calibration procedures and when power gain of applied antennas is measured. Delivered power is measured at the input of a transmitting antenna and received power at the input of a receiving antenna. In the latter case, voltage on the load may be measured instead. However, with some exceptions, at microwave frequencies, power is measured rather than voltage or current. Moreover, power is usually measured with the same (or similar) tools at both the transmitting and receiving sides. In the case considered, standard transmitting antennas are identical to standard receiving ones, and their roles are reversible; thus, we can describe a standard antenna without defining its role in a calibration set.

7.2.1   Accuracy of power measurements

In microwave metrology, power is usually measured using thermistor bridges or diode detectors. Both solutions have their advantages and disadvantages. For instance, thermistor bridges ensure better accuracy of distorted signal rms value measurement, but their dynamic range is very limited and they are sensitive to overloading. Diode detectors are more sensitive, work in a wider dynamic range, and are less sensitive to overloading; however, they are of limited application when distorted signals are to be measured. The problems are well discussed in the literature and won't be discussed here in detail. To illustrate problems related to power measurement accuracy, Table 7.2 presents set-up data for the Hewlett Packard power transducer type HP 8485A, used by the authors. The data are collected for measured power at the level of 1 mW, and the inequalities in the dynamic characteristics of the transducer, according to the manufacturer's data, are +2% to −4%. It may be noted that, occasionally, the actual value of the transducer's calibration error δ_c may exceed the declared value.

During calibrations using the transducer, the error δ_c is taken into account and correction factors are introduced to the results of calibrations performed using the transducer. The correction factors are different at different frequencies, as may be seen from Table 7.2.

Table 7.2   Correction factors of power transducer type HP 8485A for 1 mW power level

Frequency (GHz)	δ_p (%)	δ_{max} (%)	δ_c (%)	Γ_{in}
1.0	1.6	3.9		
2.0	1.6	4.2	0.2	0.006
4.0	1.7	4.3	0.9	0.003
6.0	1.8	4.5	1.5	0.013
8.0	1.9	4.6	2.0	0.026
10.0	2.0	4.8	2.0	0.036
12.0	2.0	4.9	2.7	0.03
14.0	2.2	5.6	2.7	0.019
16.0	2.3	5.5	3.8	0.037
18.0	2.0	5.6	4.9	0.042
22.0	2.1	5.7	4.3	0.069
26.5	2.1	5.3	5.5	0.043
33.0	2.4	7.1		

δ_p – the most probable transducer's calibration error,
δ_{max} – the maximal error, as declared by the manufacturer,
δ_c – individual calibration error of a device in the authors' disposal,
Γ_{in} – reflection factor at the transducers input.

7.2.2 Accuracy of transmitting antenna excitation measurement

The power delivered to a load from a receiving antenna and transmitting antenna excitation are usually measured using directional couplers. Examples of directional couplers working in the frequency range 40–112 GHz are shown in Figure 7.7 (a 9V battery indicates sizes of the couplers).

The coupler is connected between a power source and a transmitting antenna and/or a receiving antenna and its load if the power from the latter is not directly measured by a power meter.

To illustrate problems related to excitation measurement, Table 7.3 presents technical data for several directional couplers offered by Hewlett Packard.

Variations in the nominal coupling of a coupler are a function of frequency and a matching in the microwave track including the device. The variations as a function of frequency of a Hewlett Packard HP 778D type coupler, in conditions of good matching, according to the manufacturers data, are presented in Figure 7.8.

In assessing the accuracy of standard estimations, the coupling variations should be taken into account as an additional source of power measurement error. The value of this error would be equal to the maximum variation of the nominal coupling, at the level of about ±1 dB, as indicated in Table 7.3. However,

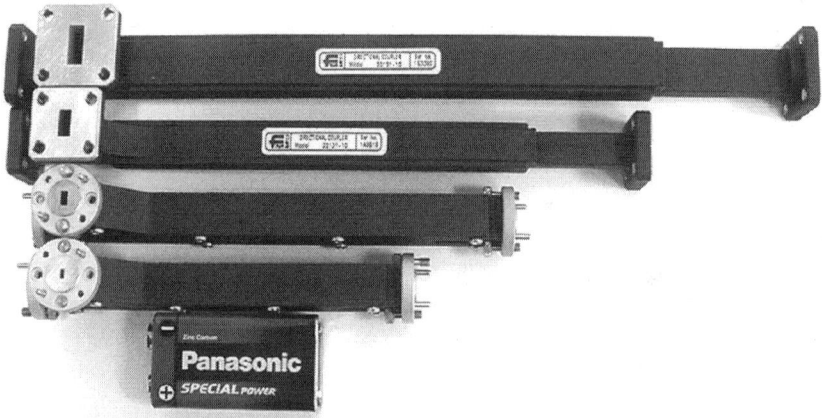

Figure 7.7 Directional couplers for frequency range 40–112 GHz

Table 7.3 Technical data of selected Hewlett Packard directional couplers

Directional couplerstype	Frequency range (GHz)	Coupling attenuation (dB)	Coupling variation (dB)	VSWR
772D	2–18	20	±0.90	1.2
778D	0.1–2	20	±1.50	1.1
779D	1.7–12.4	20 ± 0.5	±0.75	1.2
11691	2–18	22	±1.00	1.2

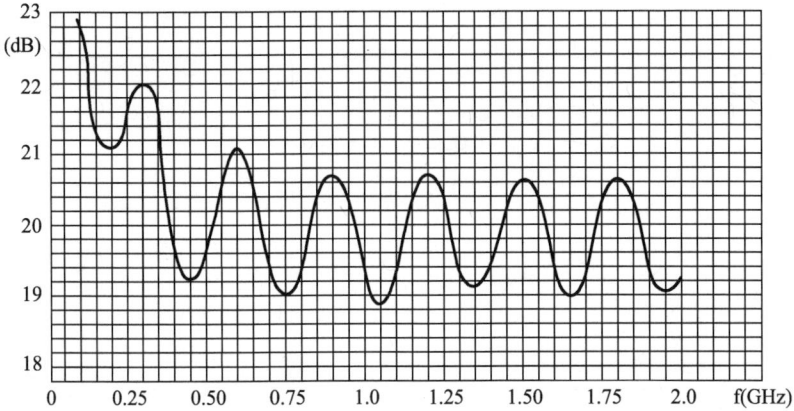

Figure 7.8   Nominal coupling variations of a HP 778D coupler versus frequency

after precise calibration of the couplers' coupling and verification of a curve as in Figure 7.8, the nominal coupling of the coupler may be estimated with an accuracy not worse than ±0.2 dB. The inaccuracy of the excitation measurement is equal to the error of the power measurement increased by the couplers' coupling inaccuracy. In our estimations of the accuracy of the standard using a standard transmitting antenna method, we will assume that the inaccuracy of the power meter, applied for exciting power measurement, is of a similar order, namely ±0.2 dB.

7.3   Accuracy of the standard estimation

Taking into account methods applied in the directional antenna calibration presented in Chapter 3.3, as well as formulas 3.9 and 3.10 that allows the calculation of the power density of a standard EMF generated both with the standard field method and the substitution method, as an example, we will present an estimation of the accuracy of standards applied by the authors.

7.3.1   Accuracy estimations of the SRA standard

The mean square error δ_r of the power density assessment when the substitution method is used is given by:

$$\delta_r = \sqrt{\delta_1^2 + \delta_3^2} \tag{7.17}$$

where: δ_1 – power measurement error,

δ_3 – power gain estimation error for the standard receiving antenna.

If we assume, on the grounds of the data presented in Table 7.2, that the maximum inaccuracy of the power measurement within the frequency range up to 10 GHz equals ±4.8%, i.e., δ_{max} for 10 GHz, and increase it to the maximal value of the error due to the nonuniformity of the dynamic characteristics of the trans-ducer, at the level of $\pm 4\%$, we will have a value for the error δ_1 not exceeding $\pm 6.3\%$.

The inaccuracy of the power gain of the SRA, on the ground of the delibera-tions presented in section 7.1, may be assumed not to exceed $\pm 5\%$. If we substitute the values of estimations of separate errors in (7.17), we will have the mean square error of the power density assessment at the level of $\pm 8\%$. The error will increase in the event of inaccurate substitution of a calibrated antenna in place of the stan-dard antenna, when a standard antenna and the calibrated one are placed axially in relation to a transmitting antenna, and if in microwave track mismatchings are not taken into account during reduction of the standard antenna gain estimations, due to reflections and multipath propagation, as a result of external interference, assuming the presence of nonlinear distortions in power exciting a transmitting antenna, as well as any other factors that could limit the accuracy of the procedure. Their presence and the role played in the standard class reduction must be estimated in each particular case.

7.3.2 Accuracy estimations of the STA standard

Applying the considerations presented above to the case of the standard field method, on the grounds of the mean square error δ_t of the standard EMF generated with the use of the STA method may be given in the form:

$$\delta_t = \sqrt{\delta_1^2 + \delta_2^2 + \delta_3^2} \tag{7.18}$$

The designations in (7.18) are identical to the case of (7.17), and δ_2 indicates here an error of the measurement of the distance between the STA and the calibrated one: $\delta_2 \leq \Delta l/l$.

Because of the application in both the SRA and in the STA of identical antennas, fully equivalent and exchangeable one to the other, the value of the error δ_3 may be assumed to be identical to (7.17). The value of the δ_3 error increases as the value of the error of the directional coupler attenuation assessment, which may be assumed to be $\pm 2\%$. Substituting the estimated values of the partial errors to (7.18) we will have $\delta_t \leq \pm 10\%$.

Similar to the case of the SRA standard, it is necessary to take into account other factors limiting the standard.

Taking into account the estimates presented, the analysis of other factors, and the results of comparative measurements, it is possible to declare that an error of the power density assessment with the use both the SRA and the STA method, applied by the authors, at frequencies up to 10 GHz, should not exceed 15%, increasing to 30% at frequencies below 40 GHz.

This estimated error of the standard may be shocking, especially if we compare it with that of the other standards of physical magnitudes, which in the case of frequency measurements may be well below 10^{-10}. However, this is the reality in electromagnetic field standards. Even the world's best EMF standards at the

National Institute of Standards and Technology are on the order of 0.5–1 dB. The gap between the accuracy of the NIST standards and the authors' is a result of the NIST's longer engagement in the field, larger staff and more experienced personnel, and better and more advanced equipment. However, our work results in continuous improvement of our standards, and any advice and suggestions in this field from the readers is appreciated in advance.

The data presented above on microwave antennas and other microwave devices does not mean that these are the best available solutions, nor are they suggested for use in the case of any standard or exposure system. The data are cited, after the manufacturers' information, in order to present typical parameters of these devices; such a set of devices has been applied by the authors in their current work.

7.4   Nonstationary EMF standard

At the very beginning of these considerations, it was assumed that any type of standard is excited from a nondistorted monochromatic sinusoidal source. Although the role of harmonics was outlined in 5.2.4.c, distortions were excluded from the presented examples of accuracy estimations. An additional problem is created by modulation of the power source applied for standardization and/or while achromatic and modulated fields are being measured. In particular, problems may exist within microwave bands. This may exist in such services as, for instance, television, cellular telephony, ISM applications, and a variety of other systems and services where a pulsed modulation is in use. At the beginning of this chapter we mentioned that thermocouple and diode detectors, which are the most popular devices applied for HF detection, have some disadvantages that limit their use. This statement was formulated in relation to pure sine wave detection. In the case of pulsed field measurements, the problem with detection increases and may lead to significant measurement errors when EMF meters designed for pulsed fields measurements are calibrated in sine wave conditions. The errors are specific to the detector type, modulation of the measured field, bandwidth, rms or peak value measurement, and other factors. Errors may be significantly reduced if a meter is calibrated in a field whose type of modulation is similar to that of the field in which the meter will be applied. The situation is quite simple when stationary fields are of concern, i.e., fields that are quasi-constant over a long time period. In the case of generator excitation, a standard should allow a required type of modulation and procedures should be described on how this may be applied.

A much worse situation appears when nonstationary fields are of concern. For instance, this could occur when a field is generated by a narrow radar beam and observed at a distance from the source. Apart from modulation with short pulses, the beam sweeps space and at the point of observation the amplitude of the pulses varies with velocity, which is a function of the rotation velocity of the radar's antenna and distance to where measurements are performed. This may lead to the conclusion that a meter calibrated even in pulsed fields may be loaded with errors of any value. Indeed, exposure time is shorter than the thermal stabilization time of a thermocouple detector. Thus, indications of a meter with a thermocouple detector will lower the real value of the EMF. Indications of a meter with a diode detector, regardless of its time constant, will depend upon the range in which the diode works (linear, square law, or between them); thus, it is not precisely known what value is being measured (rms, maximal, or something between them). In order to improve the accuracy of nonstationary field measurements, the authors proposed a nonstationary EMF standard. The nonstationary EMF may be understood as a field in a point of observation illuminated by a rotating radar beam. The radiation pattern of a radar antenna is shown in Figure 7.9.

The pattern (f_r) within the main lobe may be expressed by:

$$f_r(\varphi) = \frac{E}{E_{max}} = \cos^m \varphi \qquad (7.19)$$

where: m – a number.

Time variations of the EMF in the point of observations are presented in Figure 7.10. In Figure 7.10, A is amplitude, τ and T represent pulse modulation of a radar transmitter, T_s is the time between successive beams passing through the observation point, and τ_e reflects the beam width. The shape of the envelope of successive trains of pulses is given by (7.19). The principle of the standard is based upon the possibility of exciting an EMF standard with pulses identical to those shown in Figure 7.10.

The idea of the standard is shown in block diagram in Figure 7.11. In the set-up is a generator that allows continuous generation of pulses where the duration time (τ) may be chosen similar to that in the source; the field will then be measured with

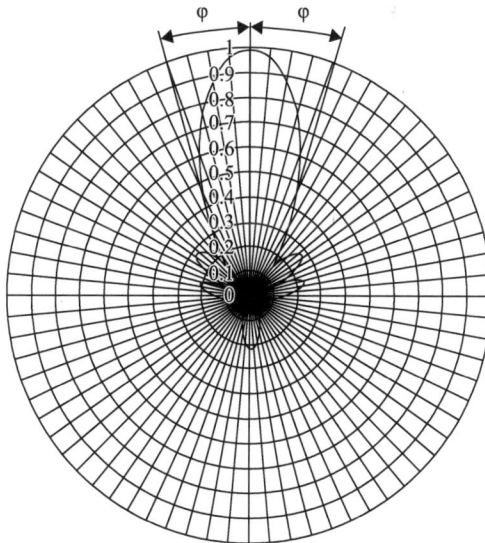

Figure 7.9   Radiation pattern of a radar antenna

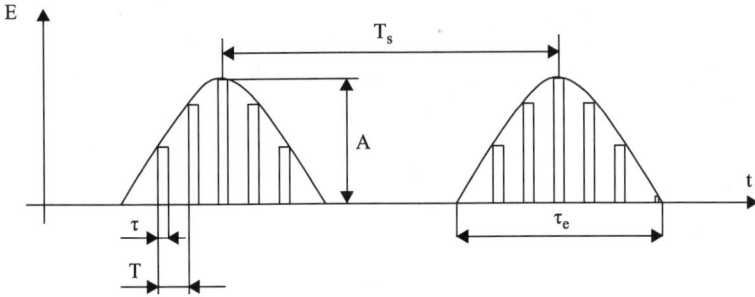

Figure 7.10   A nonstationary EMF

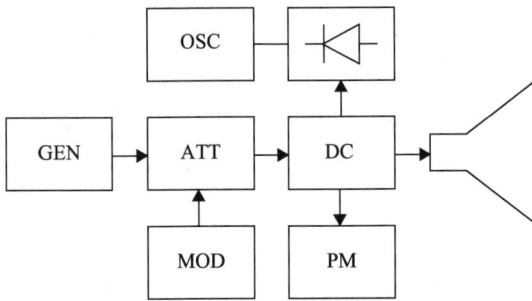

Figure 7.11   Block diagram of a nonstationary EMF standard

the use of the meter calibrated in this way. The power from the source is fed to a digital attenuator; that attenuation is modulated from an auxiliary generator. The latter generates pulses in the form of parts of a sinusoid, as given by (7.19). Duration of the pulses (τ_e) reflects time of exposure by the rotating beam, while the form of the pulses reflects the shape of the source's radiation pattern. A directional coupler makes it possible to measure the power exciting the antenna on the one hand and an observation of the generated pulses shape on the other. Meters calibrated using such a standard allow accurate measurement of nonstationary fields. Of course, the calibration should be repeated individually for each type of radiation source. Moreover, because of evident reasons, only the STA method is applicable here.

In the case of narrow beams, the factor m in (7.19) may exceed several thousands. Thus, it may be difficult to generate such pulses. In order to simplify the procedure, making it easier to use and more universal, it was assumed that the pattern presented in Figure 7.9 may be replaced by a "rectangular" one, as shown in Figure 7.12. The pattern (f_a) within the main lobe may be represented by (7.20):

$$f_a(\varphi) = \sin^2\varphi + \cos^2\varphi \tag{7.20}$$

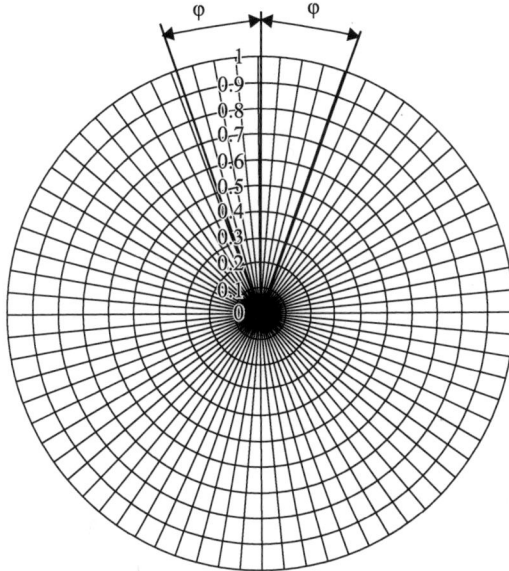

Figure 7.12   "Rectangular" radiation pattern

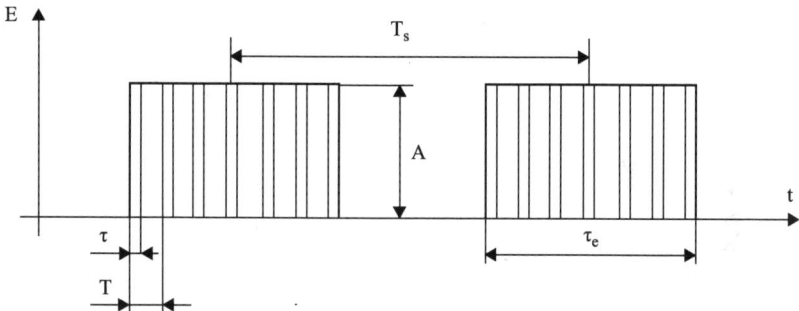

Figure 7.13   Rectangular trains of pulses

In the case considered, the EMF at a point of observation is as shown in Figure 7.13. A realization of the concept is shown in Figure 7.14. In Figure 7.14, instead of a digital attenuator (as in Figure 7.11), a key is applied while other devices may remain unchanged. Now, in order to have equivalence in calibration, the duration time of rectangular pulses (τ_e) must be reduced in relation to the previous approach. For narrow beams, the reduction is on the order of 10–15%, and it should be estimated individually for each case.

The estimations presented above were undertaken on the basis of the technical data of a source to be measured (modulation, radiation pattern, antenna rotation velocity, propagation in free space, etc.). Under real conditions, there may appear reflections, multipath propagation, presence and role of sidelobes, and other

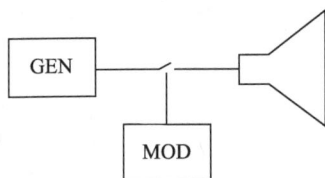

Figure 7.14   Block diagram of modified nonstationary EMF standard

sources of radiation, not to mention the difference between the real source and its data in a manual, e.g., due to aging effects. Under real conditions it is possible to measure time variations of EMF separately at any point of observation and then use the proposed method. However, this may lead to a need to perform a meter calibration not only separately for any source of radiation, but for any point of observation.

Regardless of the necessity to repeat calibrations individually for each source, the method has two advantages:

1. The use of a standard excited by a signal similar (identical) to that of the measured source allows one to neglect errors specific to the meter applied (time constant, dynamic characteristics, frequency response). This is one reason why EMF meters show differences in indications when applied to the same source measurements. The differences increase for more and more complex modulations.
2. The approach allows reduction of measuring error from an unknown to an estimatable level.

Although the method was considered in regard to nonstationary EMF measurements, a similar approach is suggested for any case of time-varying EMF.

Chapter 8

Comparative analysis of the EMF standards

Previous chapters presented and partially analyzed factors limiting the accuracy of EMF standards with dipole antennas, whip antennas, loop antennas, directional antennas, and guided waves. Several examples of accuracy estimations of the standards designed and used by the authors were introduced. Now we will try to present several methods that may be used when an estimation of separate factors limiting accuracy may be evaluated and the ways an estimation of a standard's accuracy may be verified. Although the role of the factors may be discussed without an unequivocal answer, intensive work has been done toward their full identification and evaluation. The first step here is the continuous care of the power sources, meters, directional couplers, and other auxiliary equipment. These devices must be stored in appropriate conditions that ensure their protection against electrical and mechanical failures, corrosion, and other unwanted damage that can degrade their electrical properties. Apart from this, the equipment should be subject to periodic testing and calibration. Some components, for instance, thermocouple heads or the diode detectors, may be tested and calibrated directly by their users. However, the most important devices must be calibrated in a specialized laboratory, for example, by the manufacturer or other authorized institution (as, for instance, with data shown in Tables 7.1a, 7.1b, and 7.2). Final results of the approaches, i.e., the accuracy of a standard, may be proven by a comparison with other devices. The comparisons may be performed within a single lab, with different standardization methods, or, much better, cross-comparison standards in different labs. The latter may be especially helpful in eliminating permanent errors that could neither be observed nor taken into account in a lab where they frequently appear in its routine measurements.

8.1 Double calibration method

The double calibration method was proposed by the authors years ago, and it is being used in calibrations where maximum achievable accuracy is required. The idea here is very simple; it is based upon simultaneous use of two different and independent calibration methods, for instance, the substitution method and the standard field method. The method may be applied in the calibration of dipole, whip, and loop antennas. An example of the latter application is shown in Figure 8.1. On both sides of a standard transmitting antenna, at identical distances, are placed a standard receiving antenna and an antenna of a calibrated device

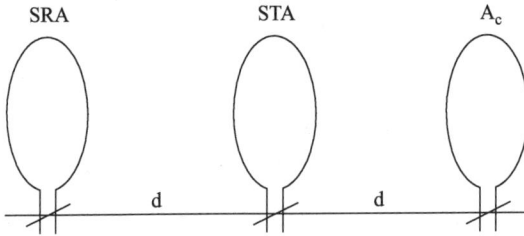

Figure 8.1   A loop antenna calibration using the double calibration method

Figure 8.2   An H-field standard using the double calibration method

(probe, sensor). After measurements are completed in the configuration as shown in the figure, the SRA and the tested antenna are replaced and the measurement is repeated. Thus, one procedure contains two calibrations done with the use of two separate and independent methods. It is expected that agreement between the obtained results should be contained within the estimated accuracy of both the standards. An example of the set-up used by the authors is shown in Figure 8.2 [20].

The method presented here is very useful when standards of similar type are compared. To illustrate the results of such a comparison, the results of calibrations performed using the world's first H-field standard working at frequencies above 30 MHz are presented in Figure 8.3 [41].

The comparison was performed with the assumption that the standard magnetic field is generated using the STA method at the level of 50 mA/m. Limits representing the estimated inaccuracy of the STA standard are indicated in the figure. The field was measured using the standard receiving antenna. As may be seen from the figure, agreement between the two methods is satisfactory within a frequency range up to 200 MHz. It may be noted that theoretical inaccuracy estimates were, in all cases, pessimistic. However, at frequencies above 200 MHz, indications of the SRA are increasingly lower, which shows that, above this frequency, there may appear factors that were not taken into account when the accuracy of the standard was estimated. Monotonic changes in the difference of the presented results would

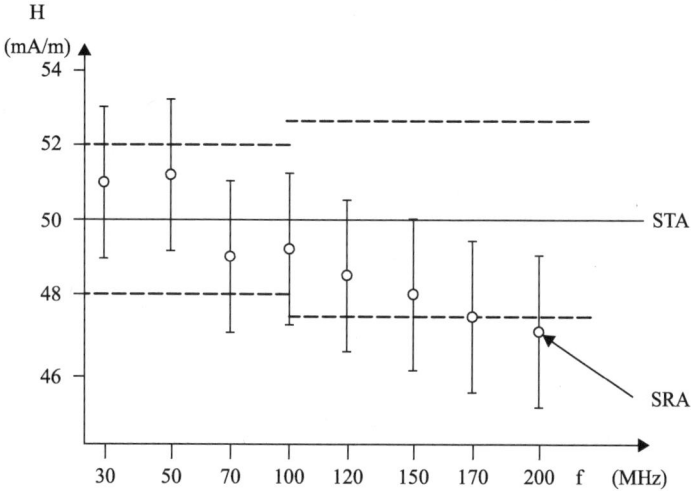

Figure 8.3 *Results of comparison of calibrations performed using the world's first H-field standard working at frequencies above 30 MHz*

suggest that the role of one factor is dominating this situation. It may be supposed that the reason for the increasing disagreement of the standards involves the electrostatic screens applied in the STA and in SRA. The antennas were as small as 5 cm in diameter, and it was necessary to screen them in order to limit couplings via the E-field that affected similar measurements performed using unscreened antennas.

8.2 Comparison of different type standards

A comparison of the standards with the double calibration method is possible only for one type of standards, as in the above case of the magnetic field standard. Apart from mutual independence of the standards, some common factors limiting their accuracy may be assumped, as with the aforementioned role of the electrostatic screen. This may lead to a compensation of errors in similar types of standards caused by similar types of the accuracy limiting factors, which results from the application of similar antennas in both standards. However, the method is irreplaceable when probes for power density measurement are tested or calibrated; this makes possible the complex and simultaneous testing of the probe in both the E-field and the H-field.

The next step in standards comparison is a comparison of different types of standards where the presence of similar factors or factors that would compensate mutually is of limited probability.

In Table 8.1, results are presented for the comparison of three standards: a TEM cell, a standard based on a standard receiving antenna, and two transfer standards: one worked out at the National Institute of Standards and Technology in Boulder, Colorado, and the other by the authors. The basic criterion of the comparison presented in the table is keeping constant the intensity of the E-field

Table 8.1   Comparison of the EMF standards

1 f (MHz)	2 E (V/m)	3 TEM (mV)	4 SRA (mV)	5 ITA TUW (mV)	6 NIST (mV)
27	2.0	3.53			3.47
	5.0	18.57			18.93
	10.0	55.82			57.35
30	2.0	3.34			3.47
	5.0	18.22			18.87
	10.0	55.50			57.33
50	2.0	3.38			3.36
	5.0	18.52			18.34
	10.0	56.15			56.10
100	2.0	3.12			3.41
	5.0	17.47			18.61
	10.0	54.63			56.50
300	2.0	2.91	2.68	2.71	2.98
	5.0	16.83		15.43	17.05
	10.0	52.95		49.12	52.59
500	2.0	2.65	2.33	2.25	3.05
	5.0	14.30	12.96	12.45	17.26
	10.0	45.85		39.80	52.95
1000	2.0		1.75	1.51	2.53
	5.0		10.56	8.14	14.66
	10.0		34.82	27.41	46.10

generated by different types of standards and then placement in the field of both the transfer standards and measurement of the voltage at their outputs. Indications in the columns of the table are as follows:

1. frequency (MHz);
2. E-field intensities at which the measurements were performed (V/m);
3. TEM: results of DC voltage measurements (in mV/m) at the output of the NIST transfer standard when the standard is placed within the TEM cell in which E-field of intensity 2, 5, and 10 V/m is generated; the mean square error of the E-field generation in the TEM cell was estimated at the level of ±5% at frequency 30 MHz and ±8% at 1000 MHz;
4. SRA: results of DC voltage measurements (in mV/m) at the output of the NIST transfer standard when the standard is placed in a standard EMF generated using the standard receiving antenna method; E-field intensities as indicated; the mean square error of the E-field generation was estimated at the level of ±8% at frequencies above 300 MHz;
5. ITA: results of DC voltage measurements (in mV/m) at the output of the NIST transfer standard when the standard was placed in EMF that was calibrated using the authors' transfer standard; the mean square error of the calibration was estimated at the level of ±10% at frequencies above 300 MHz;
6. NIST: results of measurements of the DC voltages (in mV/m) at the output of the NIST transfer standard measured at NIST at frequencies of 27–300 MHz

using a TEM cell, at frequencies of 300–1000 MHz in an anechoic chamber using a standard in the form of an open-ended waveguide, and using a standard horn antenna at 1000 MHz; as estimated by the NIST, the inaccuracy of their transfer standard calibration should not exceed ±0.8 dB.

As may be concluded from data presented in Table 8.1, agreement of the comparison of the authors' standards falls within the estimates for their classes. Agreement of our standards with the transfer standard of the NIST is good at frequencies below 300 MHz. Above 500 MHz, the difference in results of measurements exceeds the errors estimated for separate types of standards. The difference illustrates the necessity for such comparative measurements on the one hand, and may be explained by aging effects in the NIST transfer standard and effects caused by climatic factors (humidity, temperature) that were never determined for the measurements, on the other.

An example of an experimental verification of theoretical estimations is illustrated in Figure 8.4. The analysis and measurements were performed for an observation point located 3 m from a log-periodic antenna fed from a 100 W power source in the frequency range 80–1000 MHz [25].

8.3 International EMF standards comparison

In the results of EMF standards comparison presented above, performed at the EMF Standards and Metrology Lab at the Technical University of Wroclaw, a model of the NIST transfer standard, donated by the NIST to the authors for introductory checking applicability to the standards comparisons and its

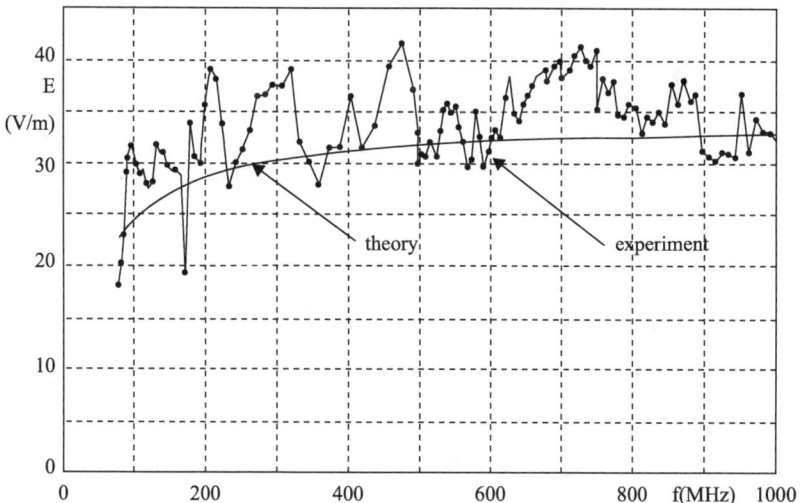

Figure 8.4   Results of comparison of theory and experiment of E-field estimates and measurements near a log-periodic antenna in the frequency range 80–1000 MHz

introductory measurements in different conditions, was applied. The NIST transfer standard was used as a reference standard in a comparison of three different standards utilized in the same lab, as presented above. This illustrates an approach to the possibility of comparison of different types of standards, designed and completed in different labs and working in different conditions, with different auxiliary equipment. On the basis of a series of comparative measurements of standards prepared in the country, it is possible to confirm that, in the majority of cases, the dispersion of measurement results remains within the frames of the declared classes of the standards compared.

Worldwide activity around the EMF standards comparison was initiated by the National Institute of Standards and Technology under the auspices of the International Bureau of Weights and Measures in Sèvres, France. The first study took place in the 1970s [40]. Results of the comparison are shown in Figure 8.5. The comparison was performed at a single frequency, 100 MHz, and was related to the standards for dipole antennas calibrated by the substitution method (SRA method). The measurements were performed in such a way that the DC output voltage of an antenna designed, completed, and then supplied for the comparisons by NIST and then delivered to the labs taking a part in the study, was between 95 and 105 mV. On the basis of the voltage, the E-field intensity was estimated equivalent to the voltage, 100 mV. Our own standard receiving antenna was calibrated in an EMF calibrated using the NIST antenna. In our case this was a dipole antenna EMF standard designed by the authors' lab and supplied for permanent use by the former Polish National Radio Inspection, another set used in EMF experiments, and calibrations by the authors. The measurements were repeated several times for different periods and over different distances between a transmitting antenna and the

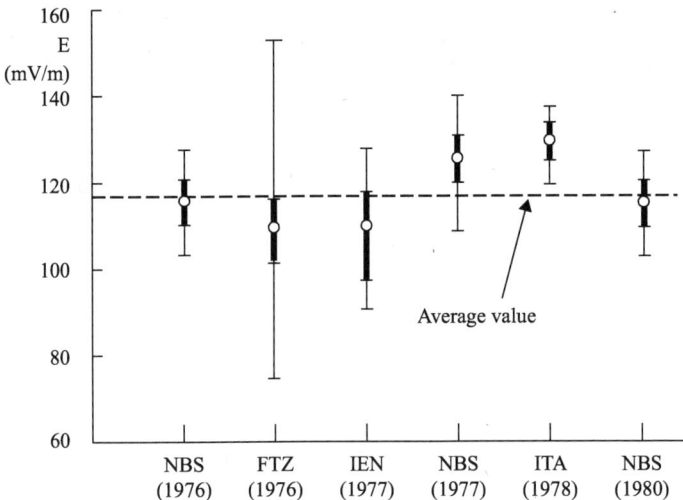

Figure 8.5   Results of international EMF standards comparison

standard receiving antennas. After completing measurements in our lab, the standard was returned to NIST. The following labs took part in the comparison:

- Instituto Electrotechnico Nazionale (IEN) in Torino, Italy,
- Fernmeldetechnisches Zentralamt (FTZ) in Darmstadt, West Germany,
- National Bureau of Standards (NBS, now NIST), United States,
- Technical University of Wroclaw (TUW), Poland.

Table 8.2 shows, in chronological order, the time and place of the performed measurements in different laboratories as well as their results. An additional factor, indicated in the table, is the number of performed measurements for the assumed configuration of the standard located at an open area test site. The number of measurements is especially important in the case of the OATS measurements and calibrations because of the impossibility of precisely controlling the temperature of the diode detector used both in the NIST transfer standard and in the calibrated ones. Detector diode characteristics are strongly dependent upon the temperature, and a larger number of measurements in different conditions allows estimation of the scale of the temperature's influence on the accuracy of the calibrations, reduction of its role, or even elimination of its influence.

Table 8.2   Detailed results of the standards comparison

Laboratory (period of measurements)	Average value of E-field (mV/m)	Number of measurements	Standard deviation (mV/m)	Estimated error (dB)
NBS (2/1976)	114.6	44	2.0	±1.0
FTZ (5/1976)	109.3	16	4.3	±3.0
IEN (5/1977)	109.7	44	6.4	±1.5
NBS (7/1977)	122.9	28	1.6	±1.0
ITA (6/1978)	125.8	16	1.6	±0.5
NBS (11/1980)	115.9	35	2.4	±1.0
Average NBS	115.7	4	1.8	±1.0

As may be seen from the data presented in Figure 8.4 and in Table 8.2, all results fall within ranges of the errors estimated by separate participants in the study. On the grounds of these results, it may be concluded that the inaccuracy of the FTZ standard estimated at the level of ±3 dB is too pessimistic, while the error of the TUW standard at the level of ±0.5 dB, in light of the results, may be supposed too optimistic.

The next international comparison, organized and headed by NIST, took place within the period 1989–1998. In the study, more labs than before were involved, namely:

1. National Physical Laboratory (NPL), England,
2. Nmi Van Swinden Laboratorium (Nmi VSL), the Netherlands,
3. Instituto Electrotechnico Nazionale (IEN), Italy,
4. Laboratoire Central des Industries Electriques (BNM), France,
5. Electrotechnical Laboratory (ETL), Japan,

6.   Korea Research Institute of Standards and Science (KRISS), Korea,
7.   Technical University of Wroclaw, Institute of Telecommunications and Acoustics (ITA), Poland,
8.   National Institute of Standards and Technology (NIST), United States.

Unlike the previous study, which was performed at a single frequency, in this study the frequency range of the comparison included frequencies from 27 MHz to 10 GHz. For comparison, three transfer standards were applied:

1.   a resonant, half wave dipole for 100 MHz;
2.   a 5 cm long dipole for frequency range 27 MHz to 1 GHz, loaded at the center with a Schottky barrier diode detector and connected to the output through a high-resistivity line; a similar model was donated to the authors and applied by them in comparisons presented in section 8.2 and shown in Figure 3.34;
3.   an 8 mm resistive loaded dipole for frequency range 0.3–10 GHz; as in the two cases above, the antenna was loaded with a Schottky barrier diode detector.

The idea of the comparison was based upon a measurement of DC voltage at the output of the antennas, using a high-input impedance voltmeter, while the antennas were exposed to three levels of sine wave EMF with a structure of linearly polarized plane waves. The levels were 0.5, 1, and 1.5 V/m for the 100 MHz dipole; 2, 5, and 10 V/m for the 5 cm dipole; and 5, 10, and 20 V/m for the 8 mm dipole. Results of measurements in separate labs were then transferred to NIST, where the final report was prepared. Selected results of measurements are presented in Tables 8.3, 8.4, and 8.5.

The tables indicate the labs participating in the comparison, the periods of time over which the measurements were performed in the lab, the inaccuracy of the standard, and measured DC voltages at the output of measured dipoles as declared by the lab. The presented data prompt several comments:

1.   The measurements were very spread out in time, usually due to problems with customs and transportation. As a result, the measurements were performed in different climatic conditions and the transfer antennas were a subject to aging effects.
2.   A disagreement of results beyond the declared inaccuracy may be seen even in results of the NIST measurements, which may suggest that participants had a tendency toward optimistic evaluation of their standards. A disagreement in relation to the averaged value could be a result of the above-mentioned reasons.
3.   Apart from NIST, the declared inaccuracy increases with frequency, which illustrates the problems faced here.
4.   Not all laboratories participated in the full range of comparative measurements, which may illustrate that even the most experienced labs might have some problems with EMF standards.

More detailed data on the study may be found in [28]. However, the publication is not free of mistakes because it was completed after the death of Prof. Kanda, the initiator and supervisor of the study.

Table 8.3 Results of comparative measurements at 100 MHz

	NIST 6/1990	ETL 6/1991	NIST 8/1993	IEN 6/1994	NIST 10/1994	KRISS 3/1995	NIST 5/1995	NPL 6/1995	NMi 8/1995	LCIE 10/1995	TUW 2/1988	Average
0.5 V/m	589 (±12%)	515 (±1.8%)	589 (±12%)		559 (±12%)	594 (±14%)	524 (±12%)	543 (±5%)	463 (±4%)	512.5 (±0.5%)	494 (±8%)	530
1.0 V/m	1223 (±12%)	1105 (±1.8%)	1186 (±12%)	1130 (±12%)	1160 (±12%)	1274 (±14%)	1140 (±12%)	1149 (±5%)	1043 (±4%)	1127 (±0.5%)	1112 (±8%)	1163
1.5 V/m	1847 (±12%)	1710 (±1.8%)	1786 (±12%)	1860 (±12%)	1760 (±12%)	1933 (±14%)	1704 (±12%)	1766 (±5%)	1633 (±4%)	1764.5 (±0.5%)	1653 (±8%)	1765

Table 8.4   Results of comparative measurements at 1000 MHz with 5 cm dipole antenna

	NIST 9/1993	NIST 11/1994	KRISS 3/1995	LCIE 11/1995	TUW 2/1998	Average
2 V/m	2.55 ($\pm$12%)	2.54 ($\pm$12%)	2.54 ($\pm$16%)	2.16 ($\pm$2%)	3.20 ($\pm$8%)	2.58
5 V/m	14.74 ($\pm$12%)	14.73 ($\pm$12%)	13.14 ($\pm$16%)	12.91 ($\pm$2%)	16.3 ($\pm$8%)	14.36
10 V/m	46.00 ($\pm$12%)	46.23 ($\pm$12%)	40.83 ($\pm$16%)	42.04 ($\pm$2%)	50.0 ($\pm$8%)	45.02

Table 8.5   Results of comparative measurements at 10 GHz with 8 mm resistive dipole

	NIST 9/1993	NIST 11/1994	KRISS 3/1995	NMi 8/1995	LCIE 11/1995	TUW 2/1998	Average
5 V/m	0.41 ($\pm$12%)	0.34 ($\pm$12%)	0.42 ($\pm$16%)	0.43 ($\pm$10%)	0.585 ($\pm$3%)	0.2 ($\pm$15%)	0.39
10 V/m	1.56 ($\pm$12%)	1.36 ($\pm$12%)	1.68 ($\pm$16%)	1.68 ($\pm$10%)	1.565 ($\pm$3%)	0.82 ($\pm$15%)	1.44
20 V/m	5.96 ($\pm$12%)	5.26 ($\pm$12%)	6.48 ($\pm$16%)	6.29 ($\pm$10%)	5.305 ($\pm$3%)	3.7 ($\pm$15%)	5.50

One general final comment summarizing the comparisons is that all the data presented show that even primary standards, in leading labs, may be of poor accuracy. Because a measuring tool cannot be more accurate than the standard used for its calibration, this illustrates what level of accuracy may be expected in offered EMF meters. If the situation with the primary standards is not very optimistic, what does that say about the accuracy of secondary standards, not to mention exposure systems? The latter seems especially important in bioelectromagnetic investigations, where the expression "accuracy of measurement" is almost forgotten.

8.4   Necessity of calibrations

Along with metrological correctness, any metrological tool requires calibration. More accurate tools, or those from which results support decision making, requires periodic testing and recalibration. This basic idea is valid in relation to EMF meters as well. Any calibration requires the use of appropriate tools, i.e., standards. The most important parameter of the standard is its class (or accuracy). Inaccurate or false estimations of the standard parameters lead to faults in the calibration of tools, even to the appearance on the market of products that do not fulfill the conditions declared by their manufacturers. Examples confirming the point are shown below.

Both the presented analysis of factors limiting accuracy of standards for EMF meter calibration and the results of comparative measurements are focused on the

Table 8.6   Results of Mild's measurements at 2.45 GHz

Producer	±0.5 dB	±1 dB	±2 dB	±3 dB
RAHAM: isotropy average value	0	18	73	91
	9	27	73	91
NARDA: isotropy average value	50	83	100	
	50	75	100	
HOLADAY: E: isotropy average value H:	22	89	100	
isotropy average value	67	78	100	
	0	0	100	
	0*	40		

necessity of preparing appropriate set-ups and methods that will ensure required accuracy. Sometimes no accuracy in itself is as important as the possibility of estimating the accuracy! If such information is not available (especially for commercial purposes), the accuracy of the standards used and calibrated can lead to disagreement between the technical data and that measured in an independent lab.

Mild has measured several series of different types of radiation hazard meters. The results of the measurements are given in Table 8.6 [6].

In Table 8.6, the numbers in the columns indicate the percentage of tested devices that fall within limits of indicated inaccuracy. The measurements were performed at frequencies of 27 MHz and 2.45 GHz and power densities within a range of 10–100 W/m^2. Mild estimated the inaccuracy of his standards at the level of ±0.7 dB. It may be seen here that the use of a standard of the class ±0.7 dB for checking meters of declared class ±0.5 dB is incorrect. Nevertheless, it may be seen from the table that several meters represent measuring errors exceeding the declared limits. The measurements at 27 MHz gave more convergent results with the manufacturers' data, but that does not change the conclusion from Mild's work. Similar measurements were performed by the authors and, based on their results, it is impossible to judge whether the disagreement of measured data and that indicated in the manuals is a result of manufacturers' faults, aging effects, overloading during their exploitation, or other phenomena. This shows the need for periodic checking of the meters, especially those used for measurements that may lead to legal decisions, as in cases of unwanted exposure, interference measurements, etc.

Figure 8.6 shows results of measurements of a frequency dependence of the correction factors of omnidirectional EMF probes. The factor reflects the ratio of the meter indications when its probe is oriented to the maximum of its pattern to that when oriented to the minimum [10]. Cook performed his measurements using eight identical EMF probes. The accuracy of his standard was estimated at the level of ±1 dB. We might repeat the comment above regarding the accuracy of the standard. However, the results indicate instability in the tested probes' parameters or problems in their manufacture.

The authors have been able to measure several types of EMF meters. Apart from the previously discussed errors in meter calibration and deformations of their

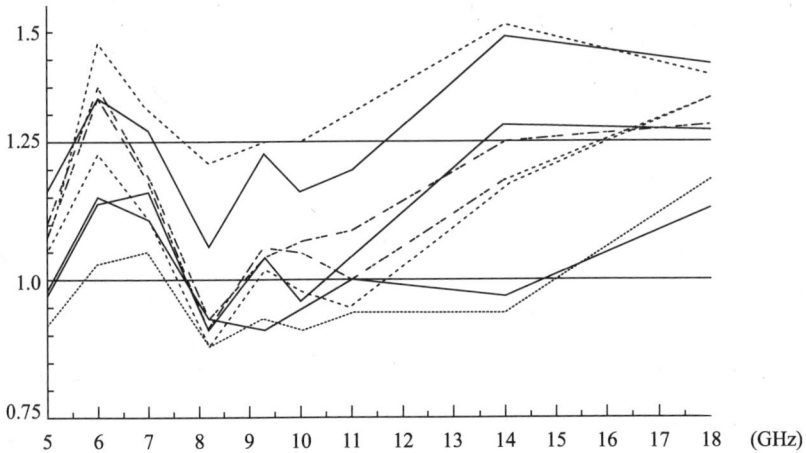

Figure 8.6   Measured frequency dependence of correction factors for omnidirectional EMF probes

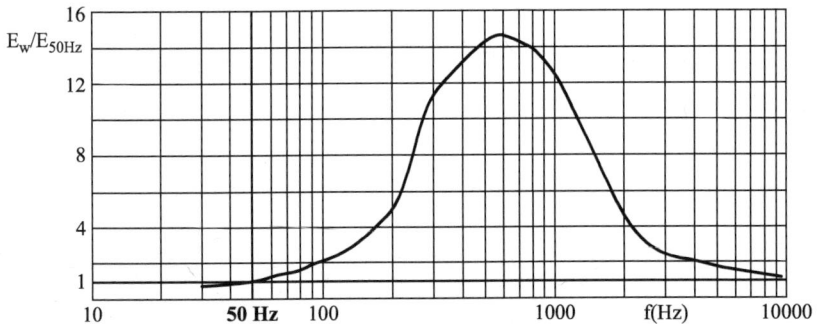

Figure 8.7   Measured frequency response of a Norad EMF meter for frequency range 50 Hz–10 kHz

directional pattern, a very important factor is the frequency response of a meter (probe, sensor). As an example, Figure 8.7 shows the relative frequency response of a NoRad EMF meter designed to work within the frequency range 50 Hz–10 kHz. As may be seen from the figure, the sensitivity of the meter at a frequency of 600 Hz is more than 10 times higher than at 50 Hz or 10 kHz. Of course, it may be said that the meter works within the declared frequency range. However, data should be consulted in which the response is presented and limitations of the meter's use discussed. It should be noticed that use of the meter for measurements of EMF at the power line frequency might lead to gross errors due to the presence of harmonics generated by thyristor power regulators, voltage converters, and other factors.

The NoRad Meter, as may be supposed from the shape of the frequency response, is a wideband meter. An example of measured frequency response of a selective magnetic field meter type MNP-89 is shown in Figure 8.8. The frequency

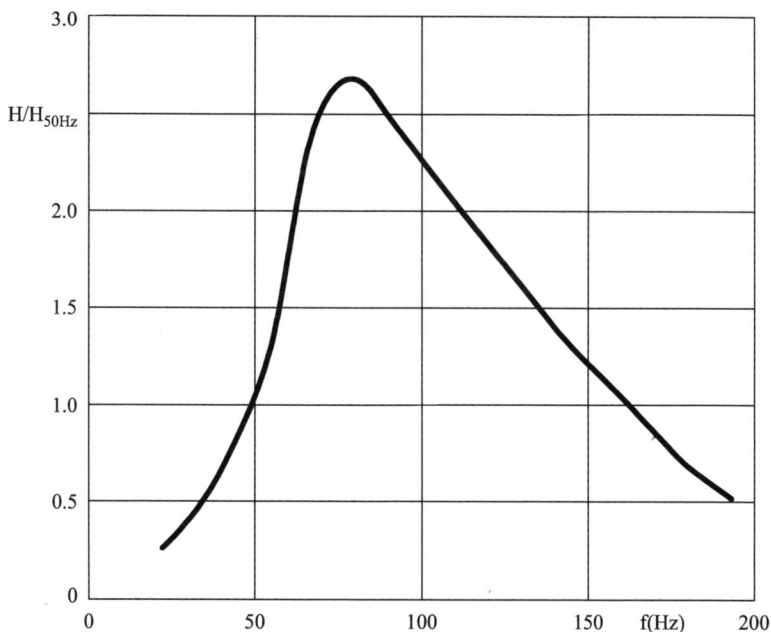

Figure 8.8 Measured frequency response of H-field meter type MNP-89 for the frequency range 45–55 Hz

range declared by the manufacturer is 45–55 Hz for sensitivity decrease of 6 dB at these frequencies, and the meter is calibrated at 50 Hz. Again, the sensitivity of the meter at the second harmonic of the power line frequency is more than two times higher as at the basic frequency.

The examples presented above clearly illustrate the need to precisely control the frequency response, and not only within the meters measuring band, but beyond as well. Figures 8.9 and 8.10 present measured frequency responses of several types of probes used by Polish surveying services. Figure 8.9 presents responses of an MH-2 magnetic field meter for the frequency range 20 Hz–2 kHz and that of probes AH-3 and AH-1 for the MEH-1 type meter, for frequency ranges 1–100 kHz and 0.1–10 MHz, respectively. Figure 9.10 shows responses of probes for the MEH-1 type meter: AE-4 for 20 Hz–2 kHz, AE-3 for 1–100 kHz, AE-1 for 0.1–300 MHz, and AE-2 for 10–300 MHz frequency ranges. All the curves are normalized in relation to the probe's sensitivity within the measuring band. The frequency response of all the probes, above their measuring band, is attenuated with the use of low bandpass filters [20,44].

Apart from the frequency response, one can often be faced with questions regarding the correctness of a device's calibration. To illustrate this, Table 8.7 shows set-up results of measurements of a PPM 8051 type meter, serial number 9961019, with probes BA 02 nr 041 and BA 05 nr 074. The measurements were performed at the EMF levels of 1, 3, and 10 V/m in the frequency range

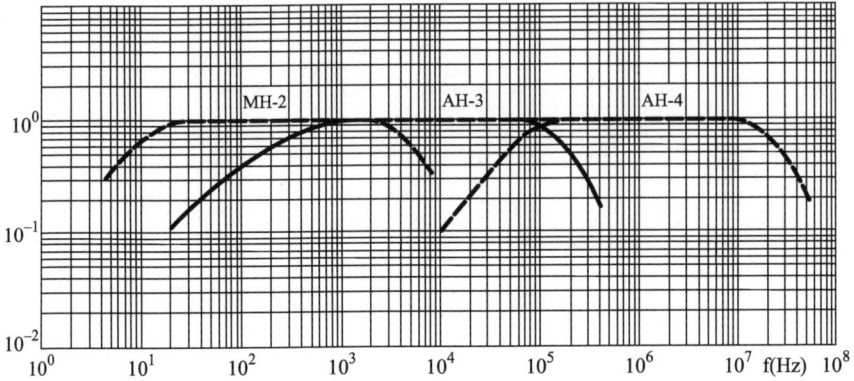

Figure 8.9   Measured normalized frequency response of magnetic field probes

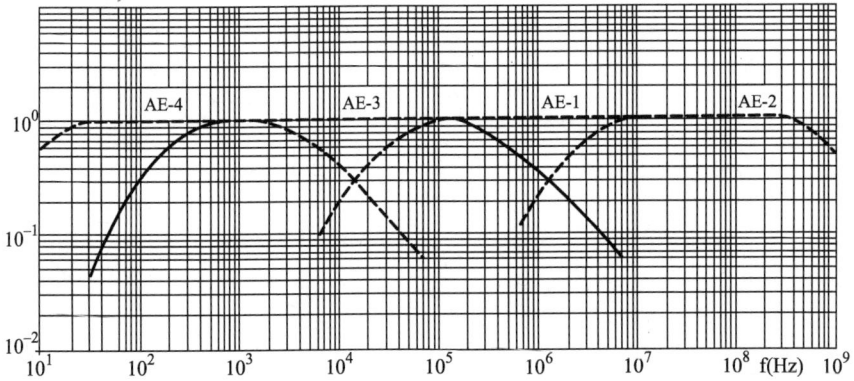

Figure 8.10   Measured normalized frequency response of electric field probes

1–1000 MHz, i.e., within the measuring frequency range and dynamic range of the meter. A TEM cell was used as an EMF standard. The inaccuracy of the standard was estimated at the level of ±8%. Similar to before, an objection may be raised about the correctness of checking a meter of declared class ±5% using a standard of class ±8%. However, the difference in the meter indications justifies such a procedure. Even if it is assumed that corner frequencies for the probes are given for the sensitivity drop of 6 dB in relation to the center of the frequency band, at the lowest frequency the attenuation exceeds 6 dB. Such a dynamic characteristic does not need any comment.

Meters tested by the authors never were new ones. They usually were sent to the authors' lab for periodic calibration, required by Polish protection standards. Thus, it is impossible to say precisely whether observed faults and imperfections in the devices were a result of damage caused through use or because of manufacturing defects. These types of meters are portable, they are usually used in heavy conditions for measurements outside the labs in industry, under different climatic

Table 8.7 Results of measurements of a PMM 8051 EMF meter

Probe BA 02 nr 041 with PMM 8051 type meter; serial nr. 9961019			Probe BA 05 nr 074 with PMM 8051 type meter; serial nr. 9961019			
f MHz	E 10 V/m	E 3 V/m	E 1V/m	f MHz	E 10 V/m	E 3 V/m	E 1V/m
1	3.0	0.0	0.0	1	2.9	0.75	0.0
3	6.0	0.0	0.0	3	3.6	1.05	0.1
5	6.0	0.0	0.0	5	3.8	1.1	0.15
10	6.0	0.0	0.0	10	3.9	1.2	0.15
20	7.0	0.0	0.0	20	4.0	1.25	0.25
26	7.0	0.0	0.0	26	4.1	1.3	0.3
30	7.0	0.0	0.0	30	4.1	1.25	0.3
40	7.0	0.0	0.0	40	4.1	1.35	0.3
50	7.0	0.0	0.0	50	4.3	1.35	0.3
100	7.0	0.0	0.0	100	4.3	1.35	0.3
150	7.0	0.0	0.0	150	4.3	1.35	0.3
200	7.0	0.0	0.0	200	4.3	1.4	0.3
250	7.0	0.0	0.0	250	4.7	1.5	0.4
300	7.0	0.0	0.0	300	5.0	1.65	0.45
350	7.0	0.0	0.0	350	5.2	1.7	0.45
400	7.0	0.0	0.0	400	5.6	1.75	0.45
450	7.0	0.0	0.0	450	5.7	1.85	0.45
500	7.0	0.0	0.0	500	5.5	1.85	0.45
550	8.0	0.0	0.0	550	5.3	1.5	0.45
600	8.0	0.0	0.0	600	5.0	1.5	0.45
650	7.0	0.0	0.0	650	5.5	1.5	0.6
700	6.0	0.0	0.0	700	6.6	1.95	0.6
800	6.0	0.0	0.0	800	5.7	1.8	0.6
900	5.0	0.0	0.0	900	5.7	1.85	0.6
1000	5.0	0.0	0.0	1000	5.4	1.8	0.6

conditions, exposed to risk of damage in more or less permanent transportation. Often they are used by personnel without appropriate experience and, therefore, for example, may be overloaded during measurements. Thermocouple detectors are particularly sensitive to this; even diode detectors may change their characteristics after overloading or overheating, especially in their nonlinear range. These factors are in addition to such things as aging effects, and many other factors. This is only one of many reasons for Poland's requirement to use a meter tester during such measurements (see section 3.5.3.2). However, the results of measurements presented by Mild, Cook, and the experience of the authors allows the conclusion to be drawn that care is not an attribute of many manufacturers, which justifies the need to check products even before their first practical application.

Something we have yet to discuss is the problem of susceptibility of EMF meters to EMF. The authors have had the opportunity to measure several types of EMF indicators, exposure indicators, smog meters, and similar devices. The situation may be summarized as "not optimistic." The majority of low-frequency devices of this type are more or less sensitive to EMF, especially at high frequencies. The effect is especially strong when there is no appropriate screening or even a plastic casing; we call them "plastic toys."

The question may be asked why the necessity of calibration is illustrated only with examples of the near-field meters. The reason is twofold:

1. The authors are mostly involved in near-field metrology, and it may be understood as their "professional deviation."

2. Calibration of meters destined for near-field metrology and accuracy of measurements are much more important compared to those associated with far-field metrology.

   - Near-field metrology is, in the majority of cases, related to labor safety and general public protection against unwanted exposure. In both cases, legal action may result if EMF levels permitted by appropriate protection standards are exceeded. In the case of professional exposure, it can lead to benefits of a financial nature (reduced working time, extra payment). In the case of environment protection, it may lead to limitations of a source activity (reduction of radiated power, reconstruction of antenna system) or even the necessity to close the source entirely.

   - Far-field measurements are usually performed using selective meters, and the role of other frequency fringes may be neglected when the near-field ones work in the majority of cases in complex electromagnetic environment that create special requirements in terms of their frequency response, susceptibility, size of probes and meters, etc.

   - Any EMF meter designated for near-field measurements may be applied to far-field measurements. Because of antennas' sizes and their directional properties, measured EMF deformations and couplings with surroundings, far-field meters are useless in near-field measurements.

   - Far-field measurements are usually performed by people experienced in electromagnetics, while surveying for exposure is very often an additional duty of people educated in chemistry, biology, or ionizing radiation and who are employed in a variety of environmental inspection services. This effect is especially important in smaller labs or local inspection services.

   - A knowledge of antennas and propagation allows the operator to foresee what may be the EMF intensity at a point of measurement in the far field and what its distribution around a point may be expected to be. In the case of the near field, such an approach is invalid. High EMF intensities may appear near low-power sources and vice versa (here is another reason for checking meters during measurements). The latter may lead to gross errors or mistakes.

   - A "human factor": The authors, performing measurements close to a variety of sources, were as a rule asked to "measure" in such a way that the levels are exceeded. The reason is obvious: the personnel exposed would like to be paid extra for work in hazardous conditions.

The differences between far-field and near-field metrology may be illustrated by results of comparative measurements presented in Table 8.8.

Table 8.8   Results of comparative measurements using different meters and antennas

Distance	E-field intensity [V/m]			
D [m]	MEH-1 + probe AE-HPE	LMZ-5 meter type + RAL-5 loop antenna H	Antenna BBH-1100 H	Antenna ADA-120 E
10	5.8	1.5	1.0	0.3
20	1.6	1.0	0.7	0.15
40	0.6	0.5	0.46	0.1
80	0.25	0.23	0.267	0.237
200	0.1	0.09	0.113	0.08

The measurements were performed at 200 kHz, near a ground plane antenna using four different meters:

(a)   a near-field meter type MEH-1 with a wideband E-field probe type AE-HP for frequency range 0.01–10 MHz,

(b)   a far-field meter type LMZ-5 with RAL-5 type loop antenna (0.15–30 MHz),

(c)   a BBH-1100 type loop antenna for frequency range 1 kHz–200 MHz, loaded with a selective microvoltmeter,

(d)   an ADA-120 type dipole antenna for frequency range 100 Hz–100 MHz, loaded with a selective microvoltmeter.

The E-field intensity was measured 1.8 m above ground level at distances (D) indicated in the table. Before measurements were taken, the meters were calibrated in different labs. The evident convergence of the results at greater distances only confirms the presence of the near-field boundary where measurement of E-field with the use of H-type antennas is not acceptable. Results presented in the fourth column indicate that the meter was probably out of order.

The above considerations lead to the following conclusions:

1.   Due to challenging working conditions, EMF meters should be the subject of periodic checking and recalibration.

2.   Because of the variety of meters available on the market for which all parameters are not always available, even newly made devices should be checked; the subject of checking should be primarily:

- sensitivity and dynamic characteristics,
- measured value (rms, peak or mean value), especially when pulsed fields or achromatic ones are to be measured,
- frequency response within the measuring band and beyond it, especially frequencies no less than the tenth harmonic of the upper corner frequency of a device,
- directional pattern of antennas (sensors) applied,
- sensitivity to nonmeasured EMF component, i.e., sensitivity of E-field meters to H-field and vice versa,
- susceptibility to EMF penetrating a device a different way than its antenna or other type of sensor,
- other parameters, depending upon the kind of planned measurements.

3. Checking of a meter may be performed using almost arbitrary tools, for instance, using a meter tester, as presented in section 3.5.3.2; the possibility of checking a meter during measurements may be profitable and is suggested.
4. Calibration of a meter, or its recalibration, should be performed by an authorized laboratory using proven standards of known accuracy.

Chapter 9

Final comments

This work briefly describes methods of standard electromagnetic field generation in terms of primary standards and different types of secondary standards including exposure systems. With some exceptions (as, for instance, exposure systems presented in section 3.5) we may summarize that there are only four approaches to standard EMF generation that are in general use, namely:

- standards with dipole antennas,
- standards with loop antennas,
- standards with horn antennas,
- standards based upon guided waves.

However, the main focus of this work has been upon the accuracy of the standards. Regardless of the standard type and its application, the factors limiting its accuracy are similar, not to say identical, as well in the case of primary standards as in the variety of their versions and modifications applied as exposure systems. The factors may be classified as follows:

1. Specificity of the method, including:

   - EMF distribution in the standard,
   - EMF homogeneity in an OUT and around it,
   - accuracy of applied formulas and their simplifications,
   - equivalence of a calibration (exposure) result to free-space conditions.

2. Accuracy of the set assembly:

   - accuracy of manufacture of separate parts of the standard,
   - accuracy of size and distance measurements,
   - accuracy of replacements (in the case of SRA method) and positioning.

3. Selection of the set excitation measurement method and accuracy of the measurement.

4. Quality of the exciting signal:

   - type and quality of modulation,
   - presence of harmonics and spectral purity,
   - presence of ripple,
   - time and frequency stability.

5. Interaction between standard and OUT:

    - EMF disturbances caused by the presence of an OUT,
    - OUT parameters affected by the standard.

6. Presence and role of "unwanted" EMF components:

    - E-field in H-field standards,
    - H-field in E-field standards,
    - susceptibility of an OUT to unwanted field components and its role in accuracy estimations.

7. Role of the standard's surroundings:

    - multipath propagation,
    - presence and role of real or artificial earth,
    - reflections and conducting caused by wiring or applied devices and other material media in the area.

8. Quality and stability of the most sensitive devices:

    - diode detectors,
    - thermistor detectors,
    - current transducers,
    - current, voltage, and power meters.

9. Presence and role of aging effects.
10. Presence and role of screens and absorbers.
11. External conditions:

    - sensitivity of a standard and an OUT to weather conditions variations,
    - presence and role of external EMI.

12. Human factors:

    - gross errors and mistakes,
    - understanding procedures to be performed,
    - experience and education,
    - equipment damage or failure,
    - misinterpretation of readings.

Although this list is quite long, it does not include all possible sources of error that should be recognized and taken into account when accuracy estimations are performed. The list is only an attempt to systematize possible sources of error in order to focus attention on selected factors that could reduce accuracy of a standard or exposure system. However, unpredicted and unexpected sources of error can always appear; several of these were discussed in this work (the role of sunlight, presence of harmonics, external EMI, mutual interactions). This leads to a very important conclusion: There are no two identical EMF standards, and even the same standard set applied in different roles and circumstances requires a reconsideration of the accuracy estimations for that particular application. A gross example here may be the role played by atmospheric conditions, especially when measurements are performed on an OATS, as well as aging effects. This could be one reason for disagreement in results of international comparison tests, even results of measurements performed at the NIST and then repeated after a period of time.

If the above is taken into account, it should be no surprise that the estimations of accuracy presented here of the same standards designed, completed, and applied by the authors may be different in different places. Moreover, the presented estimations are only examples, and nothing more, as there is no possibility to give a ready "prescription" for a standard of required accuracy. Accuracy has to be estimated individually in each case of a standard and its application.

In the presented applications, a calculation was consequently applied involving the mean square error. However, it is the opinion of the authors that, in the case of the standards, a sum of separate errors should be taken into account (maximal error) rather than the square root of the sum of squared partial errors, although such an approach is statistically less probable. Statistical manipulations can decrease the estimated value of a standard's inaccuracy, which may be profitable from a commercial point of view, but this does not mean that the standard is better. It means only that the error estimation was more optimistic. An example here could be a comparison of estimated errors shown in Table 8.2. The estimated maximal error in FTZ was ± 3 dB, but these results are close to the average value. ITA estimation looks too optimistic (± 0.5 dB), and measured results are quite far from the average value. However, the results are almost identical with those obtained at the NBS and shown in the row above that for TUW; the latter well illustrates possible problems with univocal interpretation of accuracy estimation results and reasons for agreement of calibration results (or disagreement) when performed in different labs, or even in the same lab under different conditions.

The set of standards currently applied by the authors is shown in Table 9.1. This table requires several comments:

- Decreasing accuracy of the standards with frequency increase shows both the development of the TUW standards and the metrological problems at the highest frequencies. Present activities of the lab include frequencies up to 100 GHz; however, even approximate estimation of the standards' accuracy at these frequencies is, as yet, impossible.

- When estimating the maximal values of the standards' inaccuracy, as given in the last column, only basic factors limiting the accuracy were taken into consideration. This means that, in the majority of cases, only the inaccuracy of determination of magnitudes represented in formulas describing E, H, or S in a standard were taken into account. This required an assumption that the standard is working under almost ideal conditions. Under real conditions, the inaccuracy is additionally affected by the presence and role of other factors that could appear when the standard is in use. This is one reason why the data in the table may differ in relation to that presented in other chapters.

- Maximal EMF intensities are the result of the requirements of calibration procedures and directly illustrate a standard's application. The lower ones (say up to 1 V/m) are devoted for far-field, low-level EMF meter calibration when

Table 9.1   Data of selected EMF standards designed, completed, and applied at the Technical University of Wroclaw (TUW)

Type of standard	Maximal values	Frequency range (MHz)	δ_{max} (%)
E-field, whip antennas	100 mV/m	0.01–30	±5
E-field, loop antennas	100 mV/m	0.01–30	±4
H-field, EMF sensors	1 kA/m	0–30	±10
E-field, dipole antennas	100 mV/m	30–300	±5
E-field, as above, secondary standard	100 mV/m	30–300	±7
E-field, dipole and directional antennas	1 V/m	300–1000	±8
E-field, as above, secondary standard	1 V/m	300–1000	±10
E-field, transfer standard	50 V/m	0.01–30	±5
E-field, transfer standard	50 V/m	30–300	±7
E-field, TEM cell H-field	25 kV/m 60 A/m	0–30	±5
E-field, TEM cell H-field	1 kV/m 2.5 A/m	0–1000	±8
S-field, dipole and directional antennas	50 W/m²	1000–10,000	±15
S-field, directional (surface) antennas	20 W/m²	10,000–20,000	±20
S-field, directional (surface) antennas	10 W/m²	20,000–40,000	±25
S-field, directional (surface) antennas	5 W/m²	40,000–60,000	±30

the higher EMF probe calibration as well as for EMC experiments are performed in the lab. The maximal intensities are limited only by the availability of appropriate power sources and safety precautions in order to not damage the standard.

• Although standards presented in the table may be used immediately if necessary, the table rather illustrates the area of involvement of the authors than sets that are ready for use, as the same devices and equipment are used in different combinations, for different calibrations and experiments carried out in the lab. Such an approach is not the best from the point of view of the standards themselves. They should be kept in similar conditions as the platinum standard of the meter in Sèvres. However, the approach requires precise checking of any component of the standard when the standard is set up for any particular calibration. The latter may be assumed to be an advantage of the approach, although such an approach is not advised for the reader due to mentioned rules of metrological correctness.

The discussions of the accuracy of different types of standards are related mainly to primary standards, applied as a rule, for different types of EMF meter and probe calibrations. However, their modifications and versions are mainly used in EMC investigations and in biomedical studies.

The majority of EMC investigations are done in accordance with national and international standards and recommendations. Hence, further considerations of accuracy, in a metrological sense, are not necessary apart from the accuracy of implementation and following rules and procedures provided by these regulations.

The procedures are of a commercial character, and their application is required when a product is offered on the market in order to confirm that its EM (EMI) parameters are in agreement with appropriate requirements (homologation, type testing). However, all the factors limiting accuracy must be taken into consideration when the products are prepared for manufacturing, in the design and model testing stages, not to mention many other basic EMC studies on devices and systems. Fortunately, all these measurements and experiments are performed by people with a basic knowledge in electromagnetics that ensures understanding of standardization procedures and the presence and role played by factors limiting their accuracy. However, even in the literature, there may be found examples showing that even these specialists misunderstand the problems of accuracy.

In bioelectromagnetics, in the field, the situation is still far from methodological and metrological correctness. In bioelectromagnetic studies, as a rule of interdisciplinary character, specialists representing many branches of study are involved, and the teams are mostly headed by biologists or medical doctors, while physicists or technicians play an auxiliary role. That is as it should be. The latter should not decide about the direction and scale of biomedical studies; however, the fact that they are "auxiliary" does not mean they are negligible. They should be fully responsible for designing experiments and their supervision in methodological aspects. In the field, the role of other researchers should be the auxiliary one. This is a rare case and, as a result, in the authors' estimations, more than 50% of bioelectromagnetic studies are carried out in conditions that are far from methodologically correct in regard to electromagnetics. However, interpretation of results of the studies is presented with full seriousness, and their authors require full respect to the conclusions drawn on the ground of the studies. And the conclusions may lead to very important decisions, for instance, to formulation of protection standards. Then in the standards there may be found curious solutions, from a technical point of view. For instance, many protection standards give permitted (protected) levels of exposure with an accuracy to three or more significant numbers. Not to mention that the accuracy of the levels formulated exceeds any known accuracy of biomedical studies and present bioelectromagnetic knowledge. The standards are nonrealiazable because metrological correctness requires measurement with an accuracy to the last significant figure. Imagine EMF measurements with inaccuracy at the level of 0.1% when, as may be seen from the presented considerations, the accuracy of standards, in the best case, is at the level of single percent (remember: a calibrated device cannot be more accurate than the standard applied to calibrate it!). Of course, modern digital EMF meters allow readings of as many digits as are on an indicator. Then the indications, and magnitudes recalculated from them, are presented seriously and with full respect as results of measurements, even disregarding the meter's manual, where its measuring error is usually indicated. And it is necessary to remember that years ago, exposure systems applied by Guy and Johnson represented the "state of the art" in every respect [26]. However, despite this pessimistic evaluation, there are exceptions that should be mentioned [34].

One of the most rigorous proofs of a standard is its comparison with another standard. This may be done at the individual lab level. Especially profitable are comparisons of standard EMFs generated using different methods (as mentioned in

Chapter 8), interlaboratory comparisons, and comparisons within the framework of international studies. The latter creates a good opportunity to meet specialists involved in the field of EMF standards and look deeper into the issues of accuracy. It is the authors' hope that studies initiated by NIST will be continued in the future [22]. As regards standards comparisons, it could be added that standards (exposure systems) applied in EMC investigations are usually standardized, and in that case, comparisons are not required, although they are advised. In this case, only precisely following recommended methods and procedures is required, as mentioned above. In the case of bioelectromagnetics, a variety of exposure systems is applied, and usually every lab and every team uses a set-up specially designed or adapted by themselves for research they are planning. Such an arbitrary approach is fully acceptable and profitable: It allows the construction of unique exposure systems for very specific studies. However, when results of experiments are presented, the applied exposure systems may not be discussed, which makes it impossible to evaluate the exposure estimation accuracy. This may be one reason why bioelectromagnetic experiments performed under "identical" exposure conditions in different labs lead to different results and conclusions. Again, freedom in exposure system design is doubtless an advantage; it illustrates the abilities and inventiveness of the people involved. No limitations or "standardization" is suggested here. But a precise analysis of the accuracy of such a system is indispensable. The possibility of interlaboratory comparisons of exposure systems would be very profitable. At least, a reference of an exposure to free-space conditions is suggested. These are the only conditions which are fully repeatable, and this approach is applied as a rule when any kind of EMF probes, sensors, meters, or indicators are calibrated.

Finally, a word about future trends and needs in the field. The first is the frequency range. Standards should be set in advance of any metrological applications, and there is an urgent necessity to widen the frequency range of applied standards to the highest frequencies. The second problem is related to the accuracy of the standards. It is our hope that in the near future it will be possible to generate standard EMF fields with inaccuracy below 1%. The last is a trend to automate measurements. Although many automated procedures are used, it is often done at the expense of accuracy. The most accurate standardizations still require an individual, discrete approach. Rapid development of new technologies, more and more new ideas and possibilities, as well as better and better educated people will make it possible to bring to reality needs that are, as yet, unforeseen.

The literature in the field is very extensive. A variety of presentations of designs, properties, and applications of primary standards and exposure systems may be found in the variety of papers in scientific and technical periodicals, symposia presentations, and in electronic form on the Internet. This allows one to find detailed information for any particular requirement and application. Thus, the citations refer rather to basic and general problems more than to specific ones. This approach results from the main idea of the discussion, i.e., focusing attention upon the accuracy of the standards and exposure systems, and detailed discussion of factors limiting accuracy that should be helpful in understanding the problem and its use in metrological (experimental) practice.

References

1. Altschuler H.M., Wacker P.F., Private communications, Boulder, CO 1975.
2. Babij T.M., *"The Earth Reflection Factor Measurement at Frequencies Above 100 MHz"* (in Polish), Works of IME, Techn. Univ. of Wroclaw, No. 1/1970, pp. 67–78.
3. Babij T.M., *"Detailed Requirements and Selection of a Test Site for Standard EMF Stands"* (in Polish), PhD thesis, Techn. Univ. of Wroclaw, Poland 1972.
4. Babij T.M., *"EMF Meters with Dipole Antennas Calibration"* (in Polish), Works of ITA, Techn. Univ. of Wroclaw, No. 16/1973.
5. Batchman T.E., Pavlica S., Veasey D.L., *"Amplification and Calibration for Miniature E-field Probes,"* IEEE Trans. Vol. IM-37, No. 3/1988, pp. 434–438.
6. Bostroem R., Mild K.H., Nilson G., *"Calibration of Commercial Power Density Meters at RF and Microwave Frequencies,"* IEEE Trans. Vol. IM-35, No. 2/1986, pp. 111–115.
7. Burkhardt M., Schoenborn F., Pokovic K., Kuster N., *"In Vivo Exposure Setups for Partial Body Exposure,"* Proc. 6th COST 244bis Workshop on Exposure Systems and Their Dosimetry, Zurich, Switzerland 1999, pp. 49–54.
8. Camell D.G., Larsen E.B., Cruz J.E., *"Calibration Procedures for Horizontal Antennas (25–1000 MHz),"* NBS Technical Note No. 1309, April 1987.
9. Carbonini L., *"Comparison of Analysis of a WTEM Cell with Standard TEM Cells for Generating EM Fields,"* IEEE Trans. Vol. EMC-35, No. 2/1993, pp. 255–263.
10. Cook R.J., *"EMC Measurement Standards, Calibrations and Treceability,"* Proc. 1992 Intl. EMC Symp., pp. 278–281.
11. Crawford M.L., *"Generation of Standard EM Fields Using TEM Transmission Cells,"* IEEE Trans. Vol. EMC-16, No. 4/1974, pp. 189–195.
12. Cshernomordik D.A., *"Standard Field Method for Calibration Whip Antennas within Frequency Range 0.15–30 MHz"* (in Russian), Trudy NIIR No. 3/1968, pp. 13–24.
13. Dlugosz T., Trzaska H., *"Mutual Interactions in Bioelectromagnetics,"* Environmentalist, No. 27/2007, pp. 403–409.
14. Donaldson E.E., Free R.W., Robertson D.W., Woody J.A., *"Field Measurements in an Enclosure,"* Proc. of the IEEE Vol. 66, No. 4/1978, pp. 464–472.
15. Garkavi L.H., Kwakina E.B., Ukolova M.A., *"Adaptive Reactions and an Organism Resistivity"* (in Russian), Rostov Univ. Press 1990.

16. Greene F.M., *"Influence of the Ground on the Calibration and Use of VHF Field-Intensity Meters,"* J. Res. NBS, Vol. 44, No. 4/1950, pp. 123–136.

17. Greene F.M., *"The Near-Zone Magnetic Field of a Small Circular Loop Antenna,"* J. Res. NBS, Vol. 71C, No. 4/1967, pp. 319–326.

18. Greene F.M., *"NBS Field-Strength Standards and Measurements (30 Hz to 1000 MHz),"* Proc. IEEE Vol. 55, No. 6/1967, pp. 970–981.

19. Grudzinski E., Trzaska H., *"Selected Problems of EMF Standards,"* Proc. 1992 Intl. EMC Symp., Wroclaw, Poland, pp. 265–269.

20. Grudzinski E., *"Generation and Measurements of the Standard Electromagnetic Field"* (in Polish), TUW Printing Office 1998.

21. Grudzinski E., Trzaska H., *"EMF Probes Calibration in a Waveguide,"* IEEE Trans. Vol. IM-50, No. 5/2001, pp. 1244–1247.

22. Grudzinski E., Trzaska H., *"EM-Field Standards and Their Comparison (In Memoriam of Prof. Motohisa Kanda),"* XXVII Gen. Assembly of the URSI, Maastricht 2002, p. 7.

23. Gandhi O.P., Okoniewski M., *"Computation of Electromagnetic Fields in the Human Body,"* XXVII URSI GA, Maastricht, Netherlands Aug. 2002 (CD Proc.).

24. Hansen D., Wilson P., Koenigstein D., Schaer H., *"A Broadband Alternative Chamber Based on TEM-Cell; Anechoic Chamber Hybrid Concept,"* Proc. 1989 EMC Symp., Nagoya, Vol. 1, pp. 133–137.

25. Hill D.A., Kanda M., Larsen E.B., Koepke G.H., Orr R.D., *"Generating Standard Electromagnetic Fields in the NIST Anechoic Chamber,"* NIST Technical Note 1335.

26. Johnson C.C., Guy A.W., *"Nonionizing Electromagnetic Wave Effects in Biological Materials and Systems,"* Proc. IEEE, Vol. 60/1970, pp. 692–718.

27. Kanda M., *"Standard Antennas for Electromagnetic Interference Measurements and Methods to Calibrate Them,"* IEEE Vol. EMC-36, No. 4/1994, pp. 261–273.

28. Kanda M., Camell D., de Vreede J.P.M., Achkar J., Alexander M., Borsero M., Yajima H., Chung N.S., Trzaska H., *"International Comparison GT/RF 86-1 Electric Field Strengths: 27 MHz to 10 GHz,"* IEEE Trans. Vol. EMC-42, No 2/2000, pp. 190–205.

29. Koenigstein D., Hansen D., *"A New Family of TEM Cells with Enlarged Bandwidth and Optimized Working Volume,"* Proc. 1987 EMC Symp., Zurich, pp. 127–132.

30. Koepke G., *"Measurement Procedures in Reverberation Chambers,"* Proc. 2000 IEEE Intl. Symp. on EMC, Denver, WS-3.

31. Kraus J.D., *"Electromagnetics,"* McGraw-Hill 1984.

32. Larsen E.B., *"Techniques for Producing Standard EM Fields from 10 kHz to 10 GHz for Evaluating Radiation Monitors,"* Proc. Symp. EM Fields in Biological Systems, Ottawa 1978, pp. 96–112.

33. Newell A.C., Baird R.C., Wacker P.F., *"Accurate Measurements of Antenna Gain and Polarization by an Extrapolation Technique,"* IEEE Trans. Vol. AP-21, No. 4/1973, pp. 418–431.

34. Nikoloski N., Froehlich J., Samaras T., Schuderer J., Kuster N., *"Reevaluation and Improved Design of the TEM Cell In Vitro Exposure Unit for Replication Studies,"* Bioelectromagnetics, Vol. 26/2005, pp. 215–224.

35. Omar A., Trzaska H., *"How Far Is Far?"* Applied Microwaves & Wireless, No. 3/2000, pp. 76–80.

36. Podgorski A.S., Gibson G., *"New Broadband Gigaherz Field Simulator,"* Proc. IEEE Intl. EMC Symp., Anaheim, CA 1992, pp. 435–437.

37. Podgorski A.S., *"Dual Polarization Broadband (BGF) Simulator for Emissions and Susceptibility Testing: Principles and Applications,"* Proc. 1996 Intl. EMC Symp., Wroclaw, Poland, pp. 406–408.

38. Repair A.G., Newell A.C., Baird R.C., *"Antenna Gain Measurements by an Extended Version of the NBS Extrapolation Method,"* IEEE Trans. Vol. IM-32, No. 1/1983, pp. 88–91.

39. Stratton J.A., *"Electromagnetic Theory,"* McGraw-Hill 1941.

40. Stubenrauch C.F., Galliano P.G., Babij T.M., *"International Intercomparison of Electric-Field Strength at 100 MHz,"* IEEE Trans. Vol. IM-32, No. 1/1983, pp. 235–237.

41. Trzaska H., *"Magnetic Field Standard at Frequencies Above 30 MHz,"* HEW Publication (FDA) 77-8010, pp. 68–82.

42. Trzaska H., *"Low Frequency E-field Standard,"* Proc. 1994 Intl. EMC Symp., Wroclaw, Poland, pp. 202–205.

43. Trzaska H., *"Calibration of Directional Antennas and Limitations in Their Use,"* IEEE Trans. Vol. IM-49, No. 5/2000, pp. 1112–1116.

44. Bienkowski P., Trzaska H., *"Electromagnetic Measurements in the Near Field,"* 2nd Edition, SciTech 2012.

45. Trzaska H., *"Primary and Secondary EMF Standards,"* Proc. CEEM'2006 Asia-Pacific Conf. on Environmental Electromagnetics, Dalian, China, pp. 769–770.

46. Wacker P.F., *"Theory and Numerical Techniques for Accurate Extrapolation of Near-Zone Antenna and Scattering Measurements,"* NBS Report No. 4/1972.

47. Whiteside H., King R.W.P., *"The Loop Antenna as a Probe,"* IEEE Trans. Vol. AP-12, No. 2/1964, pp. 291–297.

48. Wieckowski T.W., *"Electromagnetic Compatibility Testing of Electric and Electronic Devices"* (in Polish), TUW Printing Office, 2001.

Index

accuracy analysis 63
  of EMF standards with a segment
    of a transmission line 111,
    130–1
    homogeneity of EMF in TEM
      cell 131–5
    line accomplishment 116–17
    line excitation measurement
      accuracy 123–30
    OUT in TEM cell 135–50
    strip transmission line 112–13
    wave impedance, disturbances of
      118–23
    wave impedance of TEM cell
      113–16
  of standards with dipole antennas
    63–8
    directional antenna calibration
      85–92
    ground parameters, measurement
      of 72–4
    OATS, choice of 68–72
    SRA standard 74–81
    STA method 81–5
  of standards with horn antennas 151
    excitation measurement 159–62
    nonstationary EMF standard
      164–8
    power gain 153–9
    standard estimation 162–4
  of standards with loop antennas 93
    EMF standard with Helmholtz
      coils 104–9
    SRA standard 93–7
    STA standard 97–104
anechoic chamber 49

'antenna effect' in SRA 96–7
arbitrary structure, of EMF 7
  near and far field 7–11
  near resonant dipole 13–16
  near small loop 18–23
  of resonant dipole antenna 11–13
  of small loop antenna 17–18
  of symmetric dipole antenna
    11–13
  of thin dipole antenna 11–13

bioelectromagnetics 191

calibration methods 25
  double calibration method 169–71
calibration of meters
  with dipole antennas 25–6
    with standard field method 26–7
    with substitution method 28
    with whip antennas 29–32
  with directional antennas 39–41
  with loop antennas 32–3
    of meters with whip antennas
      37–9
    with standard field method 33–5
    with substitution method 35–7
  secondary standards and exposure
    systems 46–7
    chamber methods 47–9
    meter testers 59–61
    TEM chambers 49–54
    transfer standards 58–9
    types of 54–8
  with use of guided waves 41–2
    calibration in field of plate
      capacitor 42–3

EMF standards with traveling
    wave line 43–5
waveguide as standard EMF
    source 45–6
calibrations, necessity of 178–86
cell excitation, measuring 124
CISPR (Comité International Spécial
    des Perturbations
    Radioélectriques) 71
comparative analysis of EMF
    standards 169
  comparison of different type
    standards 171–3
  double calibration method 169–71
  international EMF standards
    comparison 173–8
  necessity of calibrations 178–86

DAMZ-4/50 log-periodic antenna 86–7
directional antenna calibration 10,
    25–6, 39–41, 85–8
  E-field averaging along the antenna
    88
  role of radiation pattern 89–92
  with standard field method 26–7
  with substitution method 28
  with whip antennas 29–32
directional antennas, calibration of
    meters with 39–41
Doppler effect 10
double calibration method 169–71

E-field 8
  inhomogeneity 132
  intensity 185
  probes 135, 140
  source 142
electromagnetic compatibility (EMC)
    investigation 1, 190–1
electromagnetic interference (EMI) 1
electromagnetic pulses (EMP) 1
electromotive force measurement
  with a thermocouple 77–9
  using a diode detector 80–1

EMCO horn antennas 155
EMF, of arbitrary structure 7
  near and far field 7–11
  near resonant dipole 13–16
  near small loop 18–23
  of small loop antenna 17–18
  of thin, symmetric, resonant dipole
    antenna 11–13
EMF meter testers (MT) 2
EMF standards with traveling wave
    line 43–5
error of calibration 15
exposure systems (ES) 2

far-field measurements 184

gross errors 67–8
ground parameters, measurement of
    72–4
GTEM chambers 51
  E- and H-field levels in 52
guided waves, meter calibration with
    41–2
  calibration in field of plate
    capacitor 42–3
  EMF standards with traveling wave
    line 43–5
  waveguide as standard EMF source
    45–6

half-wave dipoles 12, 147
Helmholtz coils, EMF standard with
    104–9
H-field 8
  antennas, by SRA method 35
  metrology 93
  probes 141
horn antennas, standards with 151
  excitation measurement, accuracy
    of 159
  power measurements 160
  transmitting antenna excitation
    measurement 161–2
  nonstationary EMF standard 164–8

power gain, accuracy of
    determination of 153–9
SRA standard estimation 162–3
STA standard estimation 163–4
HP 8485A 160
H-type antennas 185

incident voltage measurement,
    inaccuracy due to 127–30
inhomogeneity of EMF 132
input and output voltage measurement,
    inaccuracy due to 126
input voltage measurement, inaccuracy
    due to 125
international EMF standards
    comparison 173–8
International Special Committee on
    Radio Interference 71

Kuster's exposure system 55

Laplace equation 114
lightning electromagnetic pulse
    (LEMP) 1
line accomplishment, accuracy of
    116–17
line excitation measurement accuracy
    123
    inaccuracy due to incident voltage
        measurement 127–30
    inaccuracy due to input voltage
        measurement 125
    inaccuracy due to simultaneous input
        and output voltage
        measurement 126
log-periodic antenna 39
loop antennas calibration methods
    32–3
    with standard field method 33–5
    with substitution method 35–7
    with whip antennas 37–9

measurements with SRA 103–4
MEH-1 type meter 181

MH-2 magnetic field meter 181
Mild's measurements 179
MNP-89 180–1
mode reverberation chamber 48
modified Kuster's exposure system 55
monochromatic EMF 3
monopole antenna 55

National Institute of Standards and
    Technology 174
near and far field 7–11
near-field metrology 184
near resonant dipole 13–16
near small loop 18–23
NIST 192
    transfer standard 173–4
    transfer standard 58
nonstationary EMF standard 164–8
NoRad EMF meter 180
nuclear electromagnetic pulse
    (NEMP) 1

object under test (OUT) 3
    in TEM cell 135
    absorption and polarization 147
    EMF probe in TEM cell 135–42
    exposure system, TEM cell as
        142–7
open area test site (OATS) 4, 68–72

'plastic toys' 184
plate capacitor, calibration in field of
    42–3
Podgorski's open simulator 50, 51
polarization, coefficient of 3
polarization measurement, ellipse
    of 72
Polish protection standards 182
power gain measurement 153–9
PPM 8051 type meter 181

quasi-spheroidal polarization 48
quasi-spheroidal/quasi-ellipsoidal
    polarization 142

random errors 64–6
reentrant cell 53
reflection factor measurement 72–4

secondary standards 2
  and exposure systems 46–7
  chamber methods 47–9
  meter testers 59–61
  TEM chambers 49–54
  transfer standards 58–9
  types of 54–8
selective meter calibration 102–3
Siemens C35 cellular phone 55
simple antennas 7
small loop antenna 17–18
spatial orientation 3
standard EMF generation 3, 4, 25, 64
  calibration of meters, with dipole
      antennas 25–32
  with directional antennas 39–41
  with loop antennas 32–9
  meter calibration with use of guided
      waves 41–6
  secondary standards and exposure
      systems 46–61
standard field method 25, 26–7, 33–5
  dipole antennas, calibration of
      meters with 25–6
  H-field antenna calibration with 33
  loop antennas, calibration of meters
      with 33–5
standard receiving antenna (SRA) 11,
    25, 74–7, 93, 162–3
  electromotive force measurement
      with a thermocouple 77–9
  using a diode detector 80–1
  error due to nonparallel H and S_r
      vectors 95
  error due to the resonant effects
      95–6
  error due to the 'antenna effect'
      96–7
  error in e_A measurement 94
  error in frequency measurements 95

error in loop surface area
    measurement 94–5
error resulting from radii differences
    between SRA and an antenna
    under test 97
voltage measurement using a
    selective voltmeter 79–80
standard transmitting antenna (STA)
    11, 25, 81–5, 97, 163–4
  accuracy of current measurement 97
  accuracy of linear size
      measurement 97–8
  error due to noncentric placement of
      antennas 98–9
  finite sizes of calibrated antenna
      100–1
  nonlinear distortions of STA
      excitation 101–4
  nonuniform current distribution
      along STA 99–100
strip transmission line, EMF in 112–13
substitution method 25, 28–9, 35–7
  dipole antennas, calibration of
      meters with 25–6
  loop antennas, calibration of meters
      with 35–7
susceptance 119
systematic errors 66–7

TEM cell 111, 171, 182
  EMF probe in 135–42
  as exposure system 142–7
  homogeneity of EMF in 131–5
  as primary EMF standard 111
  wave impedance of 113–16
thermocouple detectors 81, 183
thin, symmetric, resonant dipole
    antenna 11–13
Thévenin's theorem 76
transmission line, segment of
  accuracy of the EMF standard with
      130–1
transmitting antenna 47
TTEM cell (triple TEM cell) 52

voltage measurement using a selective
voltmeter 79–80
voltage standing wave ratio (VSWR)
120
Volume V 8

waveguide as standard EMF source
45–6
wave impedance 118–23
calculation, mean square error of 116
of TEM cell 113–16
within disturbed area 118

whip antennas 29–32, 37–9
dipole antennas, calibration of
meters with 25–6
loop antennas, calibration of meters
with 37–9
with a standard transmitting loop
antenna 37
wideband horn 39
wideband meter calibration 103
WTEM cells (wire TEM cells) 53

Yagi-Uda antenna 39